Beginning & Intermediate Algebra

and the

TI-83/TI-84 Plus

Florence Chambers

Southern Maine Community College

Upper Saddle River, NJ 07458

Editor-in-Chief: Chris Hoag
Executive Editor: Paul Murphy
Assistant Editor: Christina Simoneau
Executive Managing Editor: Kathleen Schiaparelli
Assistant Managing Editor: Becca Richter
Production Editor: Allyson Kloss
Supplement Cover Manager: Paul Gourhan
Supplement Cover Designer: Joanne Alexandris
Manufacturing Buyer: Ilene Kahn

Pearson Prentice Hall
Pearson Education, Inc.
Upper Saddle River, NJ 07458

Printed in the United States of America

10 9 8 7 6 5 4 3 2 1

ISBN 0-13-187544-2

Pearson Education Ltd., *London*
Pearson Education Australia Pty. Ltd., *Sydney*
Pearson Education Singapore, Pte. Ltd.
Pearson Education North Asia Ltd., *Hong Kong*
Pearson Education Canada, Inc., *Toronto*
Pearson Educación de Mexico, S.A. de C.V.
Pearson Education—Japan, *Tokyo*
Pearson Education Malaysia, Pte. Ltd.

BEGINNING AND INTERMEDIATE ALGEBRA AND THE TI-83/TI-84 PLUS

BY

FLORENCE CHAMBERS

Southern Maine Community College

CHAPTER 1 GETTING STARTED

CHAPTER 2 COMPUTING WITH REAL NUMBERS

CHAPTER 3 EVALUATING ALGEBRAIC EXPRESSIONS

CHAPTER 8 LINEAR REGRESSION 51

CHAPTER 9 SYSTEMS OF LINEAR EQUATIONS 57

CHAPTER 10 POLYNOMIALS 67

CHAPTER 11 RATIONAL EXPRESSIONS AND FUNCTIONS 73

CHAPTER 12 LINEAR INEQUALITIES......79

CHAPTER 13 RADICALS AND RADICAL FUNCTIONS......91

CHAPTER 14 QUADRATIC EQUATIONS AND FUNCTIONS......99

CHAPTER 1

GETTING STARTED

THE KEYBOARD

- The keyboards of the TI-83 Plus and the TI-84 Plus are the same.
- The basic calculator functions, arrow keys, blue [2nd] and green [ALPHA] keys work in the same way.
- The blue [2nd] key accesses the blue operation or menu above a key.
- The green [ALPHA] key accesses the green character or operation above a key.
- Graphing keys are found directly below the screen.
- Screens shown in this manual are from the TI-84 PLUS Silver Edition unless otherwise noted.

Turn the Calculator On and Off

- To turn the calculator on, press the black [ON] key found at the bottom left of the calculator.
- Press [2nd] [ON] to turn the calculator off.

Contrast or Brightness

- To adjust the contrast or brightness of the screen, press the [2nd] key and the up or down arrow key, first one and then the other, until you get the desired contrast.
- [2nd] [▲] makes the screen darker.
- [2nd] [▼] makes the screen lighter.
- A number flashes in the upper right of the screen to show the level of contrast. If the number is high, 8 or 9, replace batteries as soon as possible.

The [CLEAR] Key

- The [CLEAR] key will clear the entire screen when the cursor is on a new line.
- The [CLEAR] key will also clear a single line if that line is incomplete.

Retrieve a Line

- The calculator saves several previous screens.
- To retrieve a previous line, press [2nd] [ENTER] as many times as needed to find the line you want. This action accesses the ENTRY function.

The Home Screen

- The HOME SCREEN is the primary screen of the TI-83 Plus and the TI-84 Plus, where you perform calculations.
- To return to the Home screen from any other screen, press [2nd] [MODE] to access QUIT.

Menus

The TI-83 PLUS and the TI-84 PLUS calculators have several menus. The ones that you will need the most in this course are MODE, MATH, ZOOM, and CALCULATE. Other menus you may use include the TEST, STAT, LIST, and MATRIX menus. Let's look at the MATH menu as an example of a menu. Details about the use of the items in the menus will be given later in this manual. The remaining menus will be explained as needed in this manual.

- Press the [MATH] key to view the MATH menu.
- Notice that there are four menus on the [MATH] key.
- The highlighted (darker) selection on the top line is the active menu.
- The MATH menu is the active menu shown here.
- The down arrow on 7↓fMax(means that there are more items in the menu. Press the down arrow [▾] key to view the remaining items.
- Use the right arrow [▸] key to view the other three menus.
- Use [2nd] [MODE] QUIT to leave any menu or screen.
- [CLEAR] may also get you out of some menus.

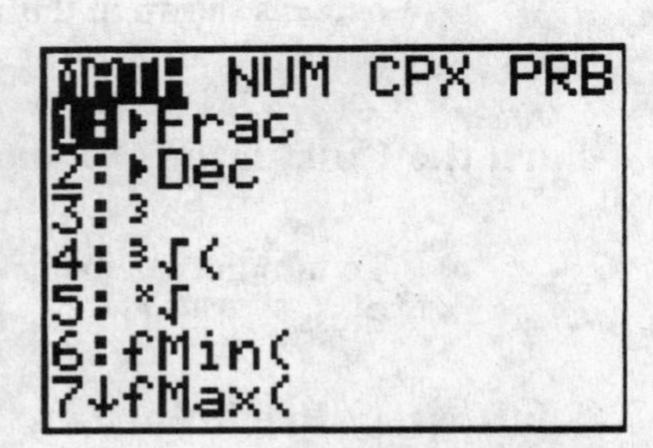

Moving from One Screen to Another Screen

- The Home Screen is the screen where calculations are performed.
- Other screens are accessed with menu keys, such as [MODE], [MATH], [Y=].
- Use [2nd] [MODE] QUIT to return to the Home screen from any other screen.
- The [CLEAR] key will return you to the Home screen from some menus such as the MODE or MATH menu.

Set the Mode

- The mode settings define how your answers will be displayed and what kind of graph you will draw, namely Function, Parametric, Polar, or Sequence.
- Press the [MODE] key to see the MODE settings.
- Use the arrow keys to place the cursor over the setting you wish to set.
- Press [ENTER] to set each mode selected.
- The selected mode will be highlighted.
- The cursor position will be blinking.
- The most common settings are discussed below.

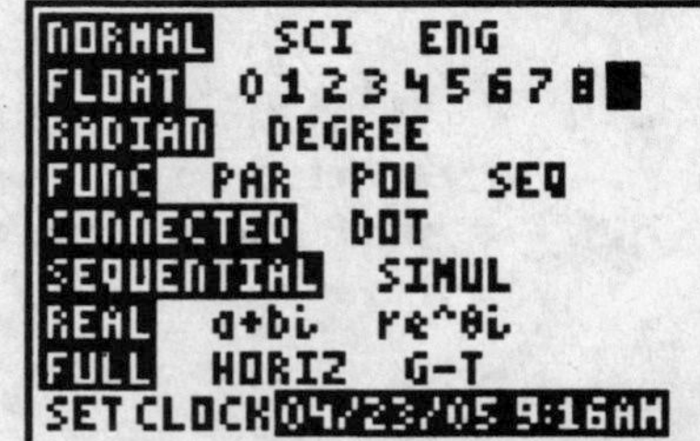

NORMAL, SCI, ENG Modes

The first line on the Mode screen defines how computations are displayed, namely, in Normal mode, scientific mode, or engineering mode.

```
253.5*439.6
          111438.6
        1.114386E5
        111.4386E3
```

- NORMAL mode is the mode for computations whose results are in normal decimal form.
- SCI mode is for computations whose results are displayed in scientific notation, as one non-zero digit to the left of the decimal point times a power of 10.
- ENG mode is for computations whose results are displayed in engineering notation. A number in engineering notation can have one, two, or three digits before the decimal. The power of 10 exponent is a multiple of three.
- The screen on the right shows a computation displayed in Normal, Scientific, and Engineering modes.

- Set this line to NORMAL.
- Use the arrow keys to place the cursor over the word NORMAL.
- Press [ENTER] to set the mode to NORMAL.
- The cursor position will be blinking.
- The word NORMAL will be highlighted when you move away from the word NORMAL.

FLOAT

FLOAT is the setting that displays the exact number of decimal places for terminating decimals, a rounded decimal for repeating decimals or irrational numbers, or a certain number of decimal places.

- For example, if you would like to have two decimal places shown when you reconcile your checkbook, set FLOAT to 2. All computations are rounded to two decimal places.

- If you would like to have four digits shown, set FLOAT to 4. All computations are rounded to four decimal places.

- Set the word "FLOAT" if you want the exact number of decimal places for terminating decimals, or a rounded decimal for repeating decimals or irrational numbers.

- Here is the number 6 shown in the Float 2, Float 4, and FLOAT settings.

```
6
              6.00
            6.0000
                 6
```

- Set this mode to FLOAT.
- Use the arrow keys to position the cursor on the word FLOAT.
- Press [ENTER] to set the mode.
- The word FLOAT will be highlighted.
- The cursor position will be blinking.
- All computations will have the exact number of decimal places for a terminating decimal, or a rounded decimal for repeating decimals or irrational numbers.
- Here are examples of three fractions that are terminating decimals, three fractions that are repeating decimals, and three irrational numbers.

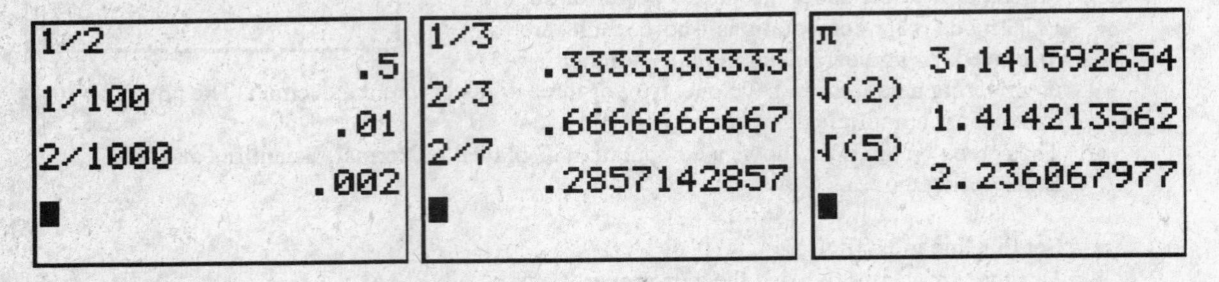

FUNC, PAR, POL, or SEQ

This line sets the kind of graph you wish to draw, namely, a function, parametric equation, polar equation, or sequence. Set this line to FUNC.

- Move the cursor over the word FUNC.
- Press [ENTER] to set the mode.
- The selected mode will be highlighted.
- The cursor position will be blinking.
- All graphs will be Y= equations.

Return to the Home Screen

- Press [CLEAR] to return to the Home screen.
- Pressing another menu key such as [Y=] or [STAT] can also get you out of the MODE screen.

Set the Clock

- The last line on the TI-84 PLUS allows you to set the date and time on the calculator.
- Select SET CLOCK and follow the screen directions.
- The settings shown on the right are a good place to start in this course.

Catalog Key [2nd] [0]

CATALOG is a complete list of all functions available on the calculator both from the keyboard and from the menus. Those functions accessed by name are listed alphabetically. Those functions accessed by their mathematical notation or symbols such as !, r, %, ≠ , are listed after the alphabetical listing. These functions may be used directly in an expression on the Home screen or other editing screen.

- Press [2nd] [0] to access the CATALOG.
- You should see a highlighted A in the upper right corner of the screen.
- If you do not see A, press the [ALPHA] key to put the calculator in alphabetic mode.
- Use the arrow keys to go to the item you want, say abs(, to use absolute value.
- Press [ENTER] to put abs(on the Home screen.
- Some functions can be put on the Y= graph screen in the same way.
- Use the up or down arrows to find another item.
- Or, you can press the first letter of the item you want if it is not on the initial screen. For example, if you want to use fMax, press [COS] to jump to the f's. Use the down arrow go to fMax, press [ENTER] to put fMax on the Home screen.

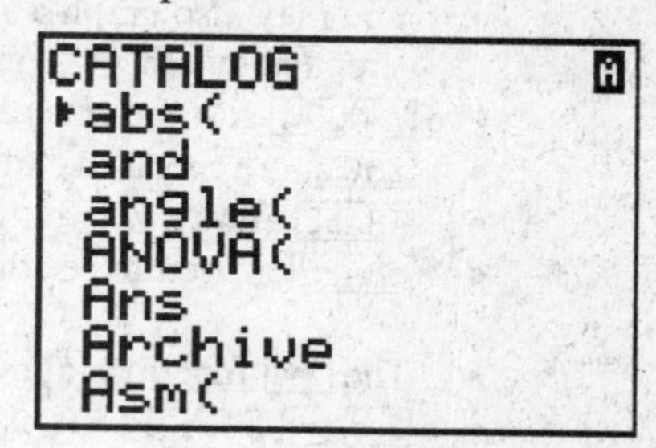

Editing Keys

The editing keys are similar to the editing keys on a computer. The arrow keys are the gray keys on the upper right of the TI-84 and the blue arrow keys on the TI-83 Plus. Insert and Delete are on the same key, [DEL].

- **Arrow keys** move the cursor around the screen.
- **Insert** [2nd] [DEL], the 2nd function on the [DEL] key, allows you to insert characters.
- Position the cursor on the character or number immediately to the right of the desired insert position. For example, if you wish to insert the number 2 between the 1 and 3 on the screen, position the cursor on the 3, press [2nd] [DEL] [2]. The 3 will be flashing. The screen will now show 123.
- **Delete** [DEL] allows you to delete characters.
- Position the cursor directly on the character or digit you wish to delete and press [DEL].
- **Enter** [ENTER] executes the expression or instruction. Press [ENTER] when you wish to see the answer or change a setting.
- **Clear** [CLEAR] removes an incomplete line on the Home Screen before you press [ENTER]. Once [ENTER] is pressed, [CLEAR] will clear the entire Home screen.

GRAPHING

Graphing Keys

Graphing on the TI-83 Plus and the TI-84 Plus is handled in the same way. There are five keys just below the screen: [Y=], [WINDOW], [ZOOM], [TRACE], and [GRAPH]. More detail about using these graphing keys will be given as they are used in this manual.

- You may input up to 10 functions in the [Y=] menu. Each function is identified by number, Y1=, Y2=, …, Y9=, Y0=.
- You may also graph a set of data points using Plot 1, Plot 2, or Plot 3 found on the graph screen above the Y= functions.
- [WINDOW] sets the size of the viewing screen.
- [ZOOM] is a menu that allows you to zoom on the graph.
- [TRACE] allows you to "walk" along the graph.
- [GRAPH] is used to see the graph of a function.

- The [2nd] functions above the graphing keys access additional graphing features.
- [2nd] [Y=] STATPLOT is used to turn plots on and off and to define the kind of plot to be drawn.
- [2nd] [WINDOW] TBLSET is used to set up a table.
- [2nd] [ZOOM] FORMAT is used to format the graph screen.
- [2nd] [TRACE] CALC is used to find and calculate important points on the graph.
- [2nd] [GRAPH] TABLE is used to display a table of values for functions that are turned on.

Each of these functions will be explained as needed in this manual.

CHAPTER 2

COMPUTING WITH REAL NUMBERS

COMPUTING ON THE HOME SCREEN

The Operation Keys [+] [−] [×] [÷]

The gray keys on the right of the calculator perform Addition [+], Subtraction [−], Multiplication [×], and Division [÷].

Add $5+2$.

- Press the [MODE] key.
- Set the Mode to NORMAL and FLOAT. The words NORMAL and FLOAT should be highlighted.
- Press [5] [+] [2].
- Press [ENTER] to see the result, 7.

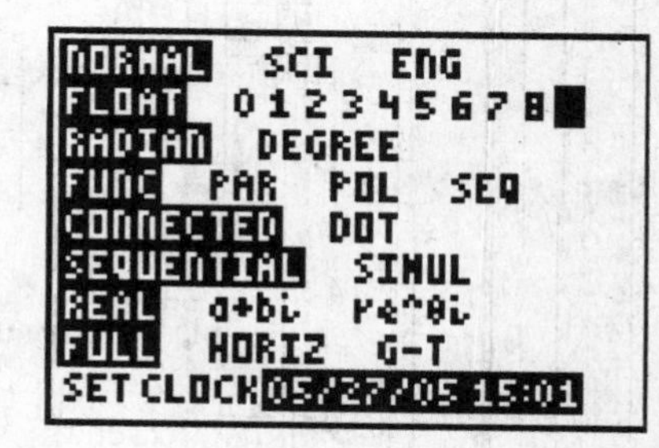

Subtract $9-3$.

- Press [9] [−] [3].
- Note that you use the subtraction key between the 9 and the 3.
- Press [ENTER] to see the result, 6.

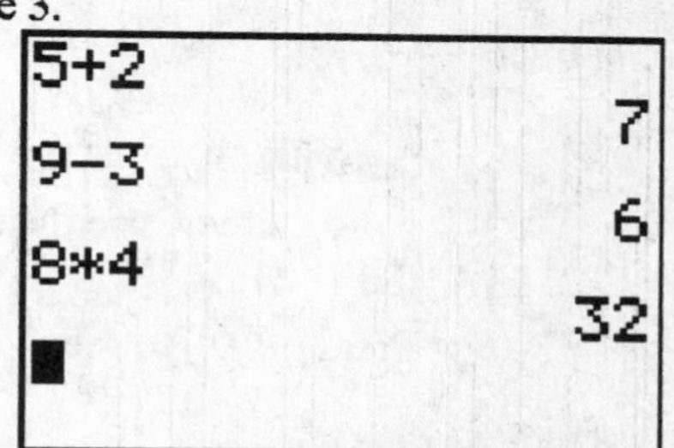

Multiply $8(4)$.

- Press [8] [×] [4].
- Note that multiplication shows up as an asterisk *.
- Press [ENTER] to see the result, 32.

Divide 25 by 5, $\dfrac{6}{2.5}$, 2/3

- Press [2] [5] [÷] [5].
- Note that division shows up as a forward slash /.
- Press [ENTER] to see the result, 5.

- Press [6] [÷] [2] [.] [4].
- Press [ENTER] to see the result, 2.4.

- Press [2] [÷] [3].
- Press [ENTER] to see the result, .6666666667.
- Note that this is repeating decimal, and the last digit is rounded.

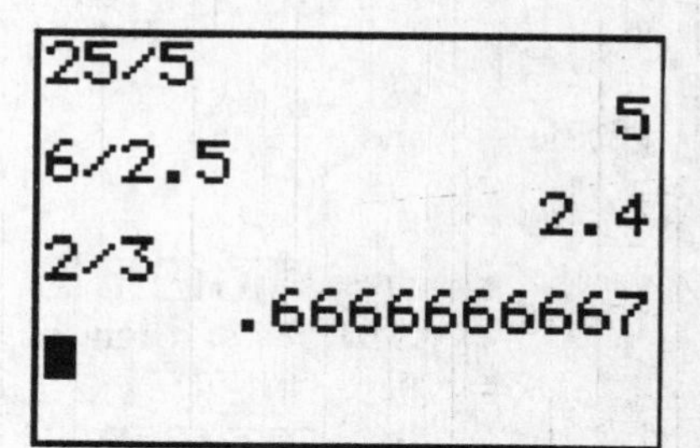

The Negation Key [(-)]

- The negation key is found in the last row of the keyboard next to the [ENTER] key.
- Note that the negation key has parentheses around the minus sign.
- The negation key [(-)] *MUST* be used when the first term of an expression is a negative number.
- The negation key *MUST* be used when multiplying or dividing with a negative number.
- The negation symbol shows up shorter and higher on the screen than the subtraction symbol.
- The negation key *CANNOT* be used between terms. If you use [(-)] between terms, you get as ERR: SYNTAX screen.

Computing with Negative Numbers

Add $-5+2$.

- Press [(-)] [5] [+] [2].
- Press [ENTER] to see the result, -3.

Subtract $-9-3$.

- Press [(-)] [9] [−] [3].
- Note that you use the subtraction key between the 9 and the 3.
- Press [ENTER] to see the result, -12.

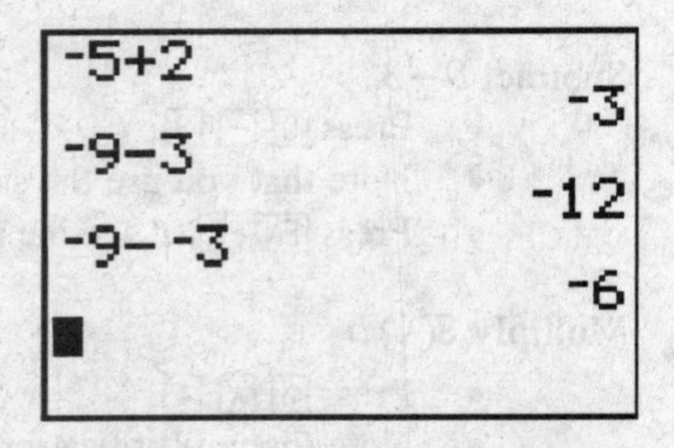

Subtract $-9-(-3)$.

- Press [(-)] [9] [−] [(-)] [3].
- Note that you use the subtraction key between the 9 and the (-3).
- Press [ENTER] to see the result, -6.

Multiply $-8(4)$.

- Press [(-)][8] [×] [4].
- Note that multiplication shows up as an asterisk *.
- Press [ENTER] to see the result, -32.

Divide $\frac{25}{-5}$ and $\frac{-6}{-2}$.

- Press [2] [5] [÷] [(-)] [5].
- Note that division shows up as a forward slash /.
- Press [ENTER] to see the result, -5.
- Press [(-)][6] [÷] [(-)] [2].
- Note that division shows up as a forward slash /.
- Press [ENTER] to see the result, 3.

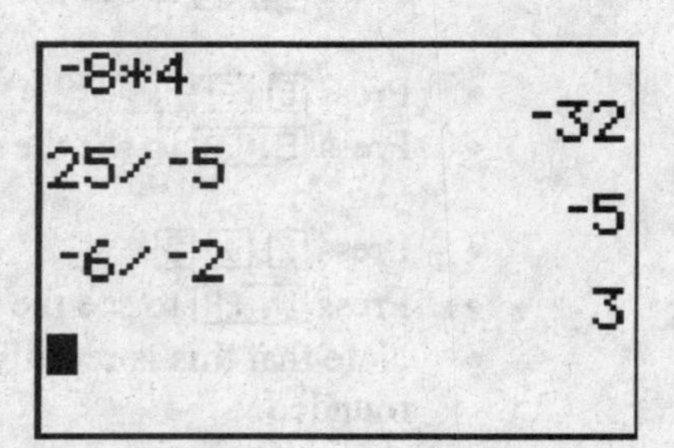

Use (-) the negation key [(-)] when the first term of an expression is negative

Evaluate $-5+3(4)$.

- Enter [(-)] [5] [+] [3] [×] [4].
- Press [ENTER] to see the result, 7.
- The negation key is to the left of the [ENTER] key.
- The negation sign shows up smaller and higher than the subtraction sign.

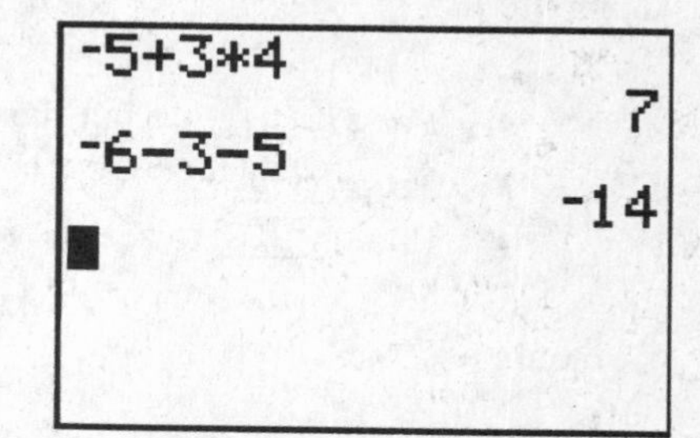

Use the subtraction key [−] between terms

Evaluate $-6-3-5$.

- Enter [(-)] [6] [−] [3] [−] [5].
- Press [ENTER] to see the result, -14.

Squaring a value using [x^2] or [^]

Squaring can be entered either with the [x^2] key or with the power key. The screen looks different for each one, but the computation is the same.

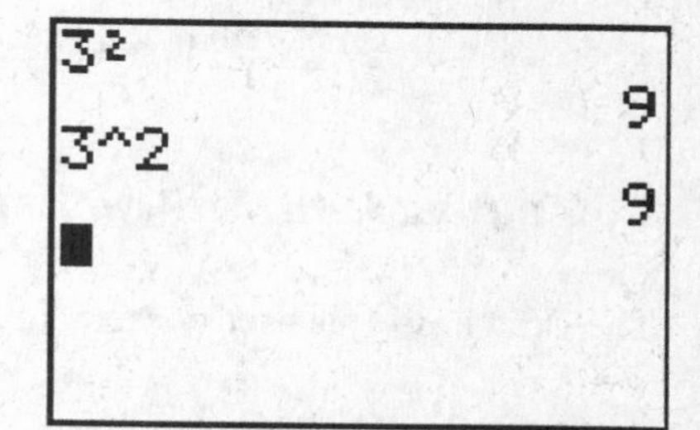

Compute 3^2.

- To use the [x^2] key, enter the number 3 first, then press [x^2].
- Press [ENTER] to see the result.
- To use the [^] key, enter the number 3, then press [^] [2].
- Press [ENTER] to see the result, 9.

Squaring with Negatives

Compute -3^2 and $(-3)^2$.

- When you negate 3^2, you write -3^2, which means $-(3 \cdot 3)$.
- To compute -3^2 on the calculator, enter [(-)] [3] [x^2].
- Press [ENTER] to see the result, -9.
- When the base is a negative, say -3, you write $(-3)^2$ which means $(-3)(-3)$.
- To compute $(-3)^2$ on the calculator, enter [(] [(-)] [3] [)] [x^2].
- Press [ENTER] to see the result, 9.

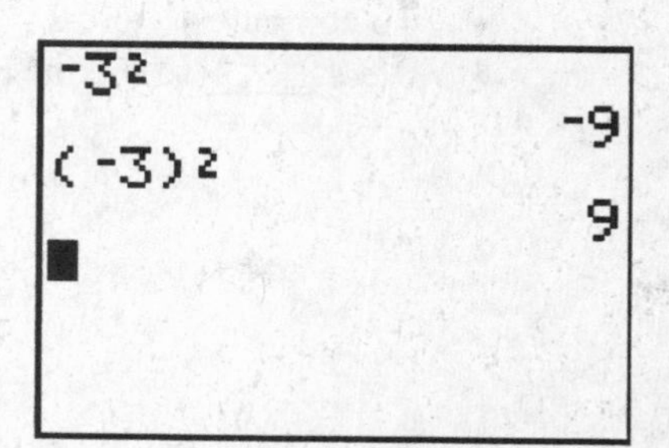

Square Root

Square root $\sqrt{\ }$ is found above the [x^2] key.

Compute $\sqrt{4}$.

- Press [2nd] [x^2] to put the square root symbol $\sqrt{\ }$(on the screen.
- Press [4] [)].
- Press [ENTER] to see the result, 2.

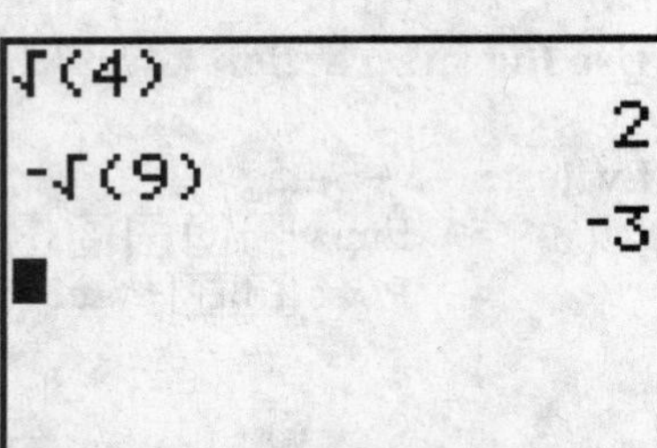

Compute $-\sqrt{9}$.

- Press [(-)] to put the negative sign on the screen.
- Press [2nd] [x^2] to put the square root symbol $\sqrt{\ }$(on the screen.
- Press [9] [)].
- Press [ENTER] to see the result, -3.

Computing with Positive Exponents using [^]

The carat key [^] is used for exponents. The [^] key is a black key found in the fourth row of keyboard.

Compute 2^3, 3^4, 4^5.

- To compute 2^3, press [2] [^] [3].
- Press [ENTER] to see the new result, 8.

- To compute 3^4, press [3] [^] [4].
- Press [ENTER] to see the new result, 81.

- To compute 4^5, press [4] [^] [5]
- Press [ENTER] to see the new result, 1024.

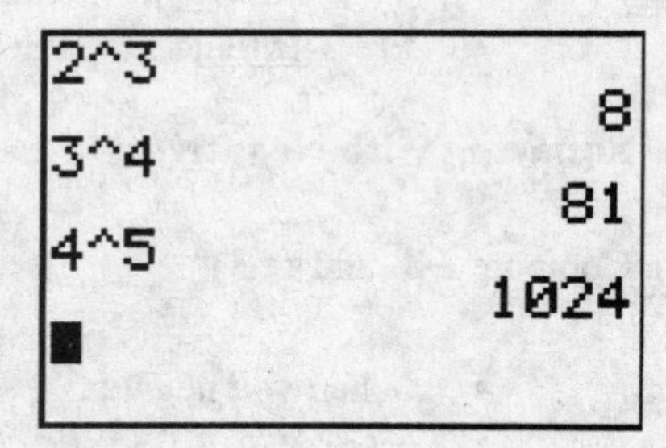

Computing with Negative Exponents using [^]

The carat key [^] is used for exponents. Negative exponents are defined by the rule, $a^{-n} = \frac{1}{a^n}$.

Compute 2^{-3}, $\frac{1}{2^3}$.

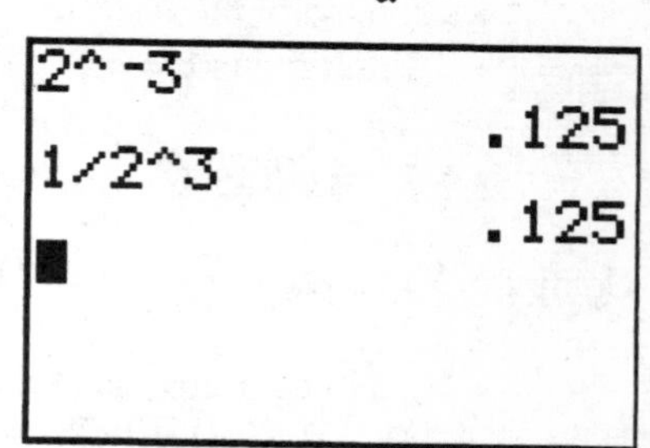

- To compute 2^{-3}, press [2] [^] [(-)] [3].
- Press [ENTER] to see the result, .125.

- To compute $\frac{1}{2^3}$, press [1] [÷] [2] [^] [3].
- Press [ENTER] to see the result, .125.
- Note that the results are the same value, .125.

Using the answer ANS in a computation

Every time you press [ENTER] on the Home Screen, the answer is automatically stored in the calculator as ANS. Note that *ONLY THE LAST* result is stored in ANS. There are two ways to use ANS, by pressing an operation key [+] [−] [×] [÷] first on a new line, or by pressing [2nd] [(-)] to use ANS anywhere in the computation.

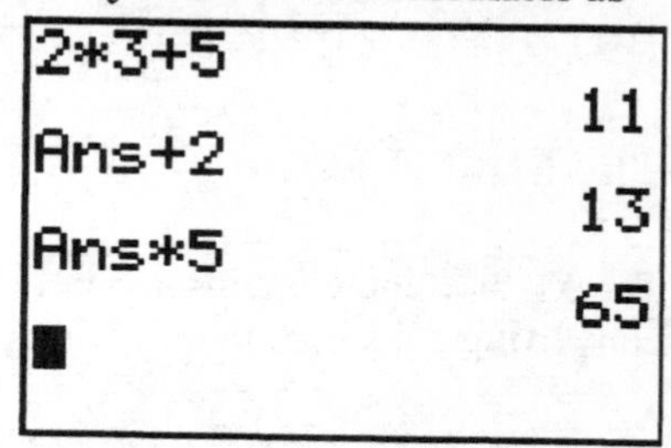

ANS using an operation key [+] [−] [×] [÷] on a new line

- When an operation key [+] [−] [×] [÷] is pressed as the first item on a new line, the calculator uses the last computed value that it stored in the variable ANS in the next computation.
- You will see ANS+, ANS-, ANS*, or ANS/ first, depending upon the operation used.
- Enter the expression $2*3+5$ on the Home screen.
- Press [ENTER] to see the result, 11.
- Enter [+][2]. The screen shows ANS +2.
- Press [ENTER] to see the new result, 13.
- Enter [×][5]. The screen shows ANS *5
- Press [ENTER] to see the new result, 65.

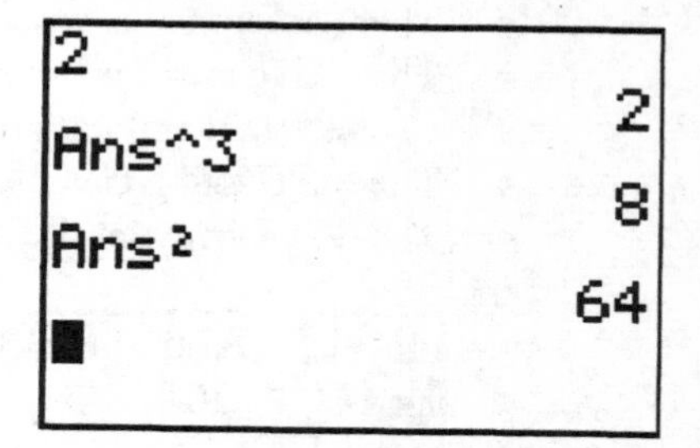

ANS using other keys [^] [x^2] [x^{-1}] on a new line

- Enter [2].
- Compute ANS^3, ANS^2, ANS^{-1} as shown on the right.

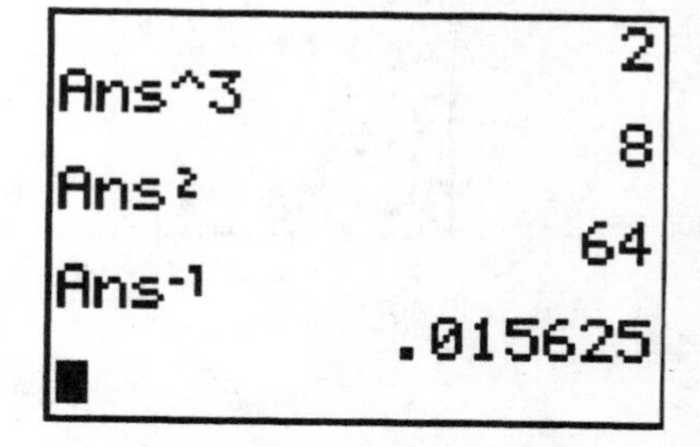

Using *ANS* [2nd] [(-)] anywhere in the expression

- The variable ANS is found as a second function of the negation key, [2nd] [(-)].
- Enter the expression, $19-2$ as shown.
- The result, 17, is now stored in the variable ANS.
- Use [2nd] [(-)] (ANS) wherever you wish to use it in the next computation.

Compute $5+ANS$

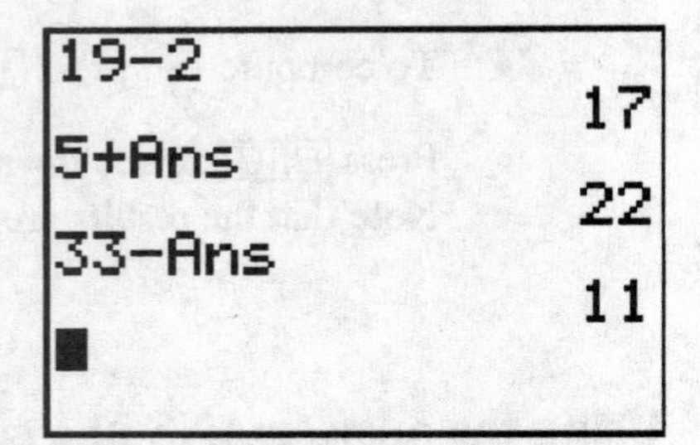

- ANS contains the number 17.
- Press [5] [+] [2nd] [(-)].
- Press [ENTER] to see the result, 22.

Compute $33-ANS$.

- ANS contains the number 22.
- On the new line enter [3] [3] [−] [2nd] [(-)].
- Press [ENTER] to see the new result, 11.

The MATH Menu

The Math Menu contains a variety of functions of interest in this course, such as absolute value, converting a decimal to a fraction, cube root, and nth root.

- Press [MATH] to access the Math Menu.
- There are 4 menus within the Math Menu, MTH, NUM, CPX, and PRB.
- The active menu is highlighted.
- Use the right and left arrow keys to move to another menu within the Math Menu.
- The MTH and NUM menus are used the most in this course.
- The down arrow at the bottom of the screen means that there are more items in the menu.

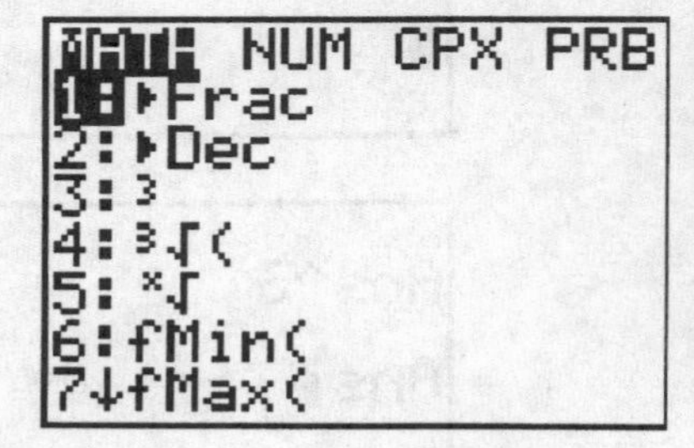

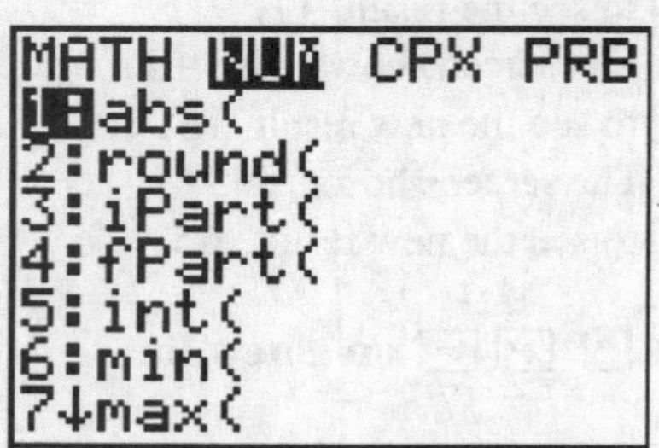

Use →FRAC to convert a decimal to a fraction

- To convert the decimal 0.25 to a fraction, first enter 0.25 on the Home screen.
- Next, press the [MATH] key.
- The cursor is on 1: →FRAC.
- Press [ENTER] or the [1] key to put →FRAC on the Home screen
- Press [ENTER] again to see the result, 1/4 .

0.25

MATH NUM CPX PRB
1:▸Frac
2:▸Dec
3:³
4:³√(
5:ˣ√
6:fMin(
7↓fMax(

0.25▸Frac
1/4

Use →FRAC to convert a repeating or terminating decimal to a fraction

Add $\frac{2}{3}+\frac{5}{4}$. Convert the result to a fraction.

- When we add a fraction to a fraction we get a fraction.
- Enter $\frac{2}{3}+\frac{5}{4}$ on the Home screen.
- Press [2] [÷] [3] [+] [5] [÷] [4]
- Press [ENTER] to see the result, 1.916666667 .

- To convert 1.916666667 to a fraction, press the [MATH] key.
- Since 1: →FRAC is the first item in the highlighted MATH menu, press [ENTER], or press [1]. This puts →FRAC on the Home screen.
- Press [ENTER] again to see 23/12, the fractional equivalent of the answer, $\frac{23}{12}$.
- Note that you can write this all on one line as shown below.

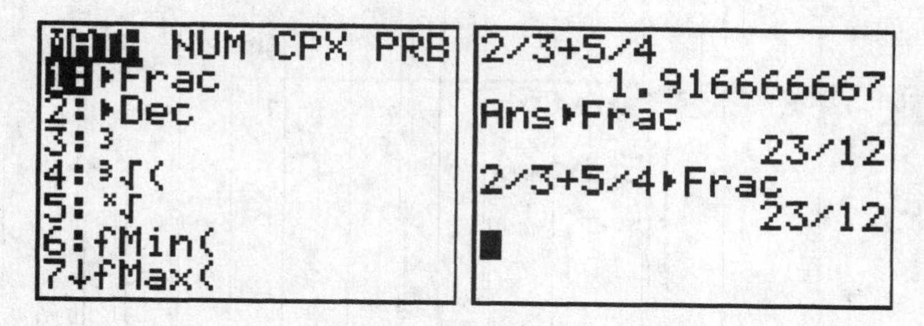

Chapter 2: Computing with Real Numbers

Use absolute value abs

The abs(is a grouping symbol. It evaluates the absolute value of a number or an expression.

Evaluate $|-5|$.

- Press the [MATH] key.
- Use the right arrow to move to the NUM menu.
- The cursor is on 1:abs(.
- Press [ENTER] or the [1] key to put abs(on the Home screen.
- Press the negation key [(-)] [5] to enter the value -5.
- Press [)] to close the parentheses.
- Press [ENTER] to see the result, 5.

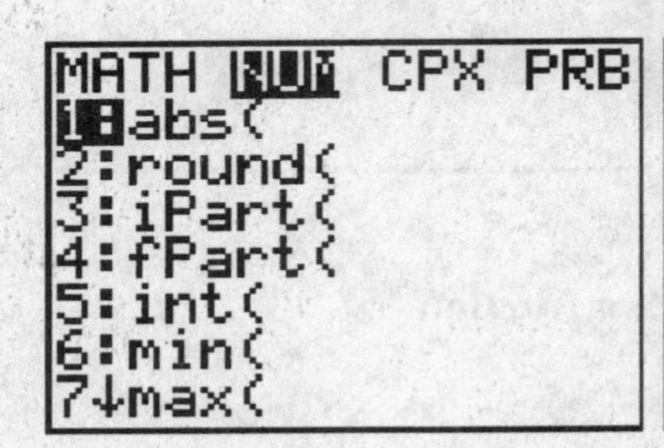

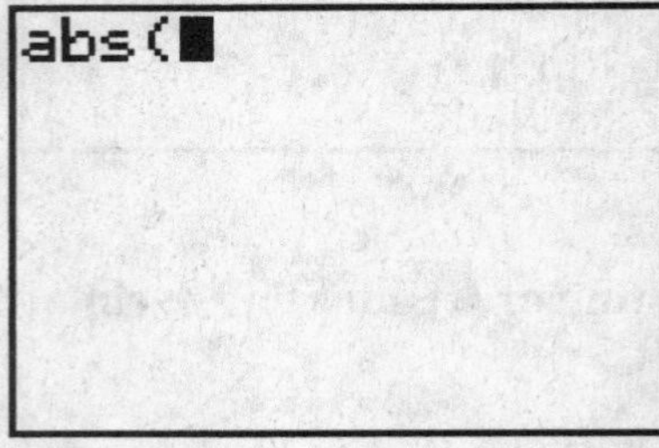

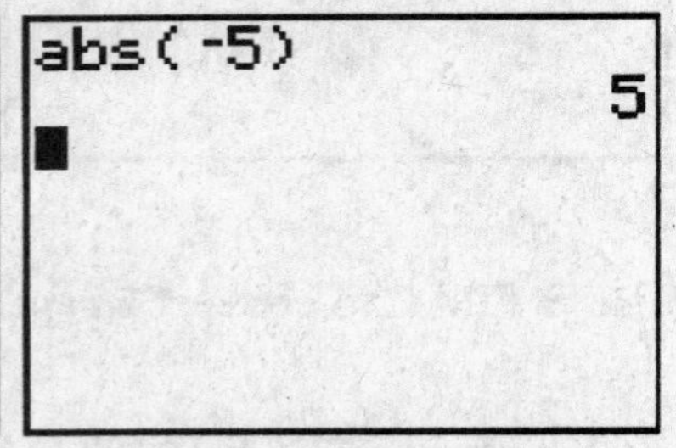

Evaluate $|5-7|$.

- To evaluate $|5-7|$, press the [MATH] key.
- Use the right arrow key [▶] to move to the NUM menu. abs(is the first item in the highlighted NUM menu.
- The cursor is on 1: abs(.
- Press [ENTER] or [1] to put abs(on the Home screen.
- Enter the expression $5-7$.
- Press [)] to close the parentheses.
- Press [ENTER] to see the result, 2.

Evaluate $3|6-9|$.

- Press [3] [MATH] [▶] [ENTER] to put 3 times abs(on the Homes screen.
- Press [6] [−] [9] [)].
- Press [ENTER] to see the result, 9

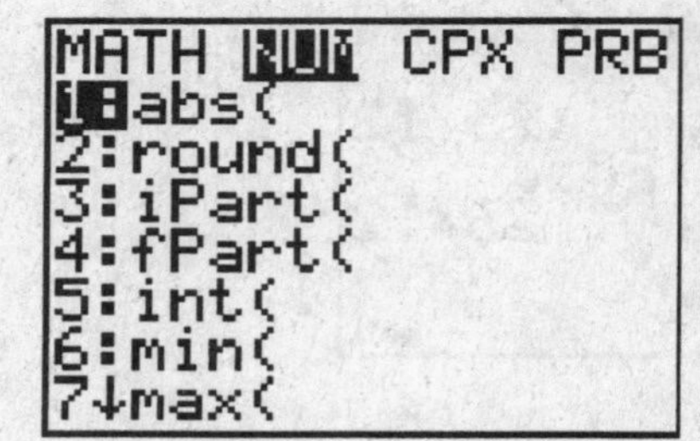

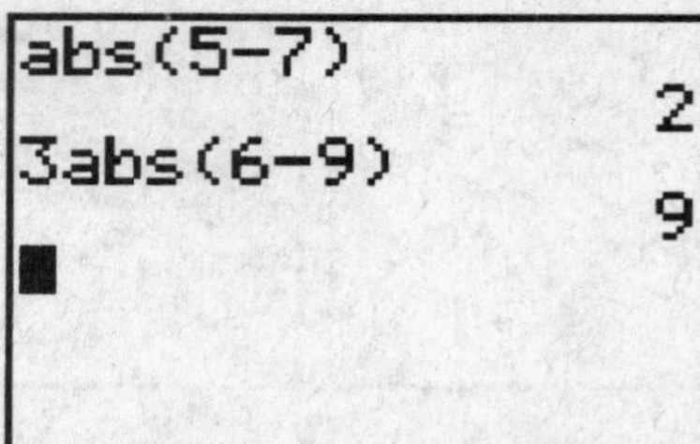

Some Uses of Parentheses

Use parentheses when there is more than one term in a radical or absolute value, and when the numerator or denominator of a fraction has more than one term.

Evaluate $\sqrt{(9+5)}$ as shown.

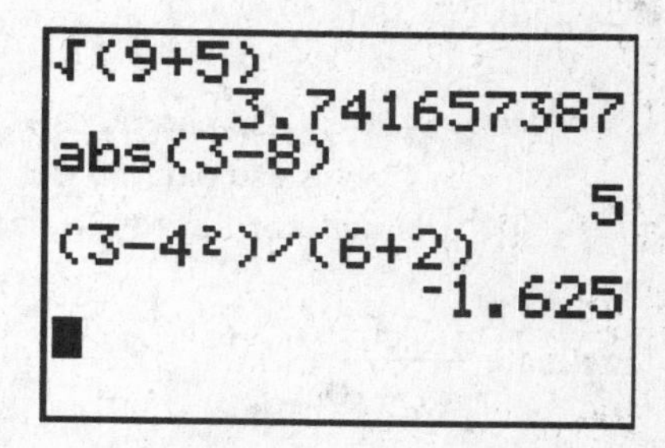

- Press [2nd] [x^2] [(] [9] [+] [5] [)].
- Press [ENTER] to see the result, 3.741657387.

Evaluate $|3-8|$.

- Press [MATH] [▶] [ENTER] to put abs(on the Homes screen.
- Press [3] [−] [8] [)].
- Press [ENTER] to see the result, 5.

Evaluate $\frac{3-4^2}{6+2}$.

- Press [(] [3] [−] [4] [x^2] [)] [÷] [(] [6] [+] [2] [)].
- Press [ENTER] to see the result, -1.625.

Implied multiplication

Implied multiplication means that the calculator understands that parentheses can mean multiplication.

Evaluate $5(1+2)$.

```
5(1+2)
                15
-4(-3-5)-3(8-2)
                14
```

- Enter [5] [(] [1] [+] [2] [)].
- Press [ENTER] to see the result, 15.

Evaluate $-4(-3-5)-3(8-2)$.

- Press [(-)] [4] [(] [(-)] [3] [−] [5] [)] [−] [3] [(] [8] [−] [2] [)].
- Press [ENTER] to see the result, 14.
- Note that the negation key *must* be used in the first expression before the 4, *and* before the 3.
- The negation sign shows up smaller and higher than the subtraction sign on the calculator.

Parentheses, Order of Operations, and Grouping Symbols

Grouping symbols are always performed first. Common grouping symbols in mathematics are parentheses, brackets, and braces. Absolute value symbols, and radical symbols are also considered grouping symbols.

- *Always* use parentheses for computations on the graphing calculator.
- *Always* use parentheses within absolute value and radical symbols.
- Brackets refer to matrices, and braces refer to lists.

Evaluate $\frac{(-3)(-4)}{7-10}$.

- When a fraction has more than one factor or term in the numerator and/or the denominator, you must enclose the numerator in parentheses and the denominator in parentheses.
- Press [(] [(-)] [3] [)] [(] [(-)] [4] [)] [÷] [(] [7] [−] [1] [0] [)].
- Press [ENTER] to see the result, -4.

Evaluate $(6-8)^2(5-7)^3$.

- When the sum or difference of terms is raised to a power, you must enclose the sum and difference in parentheses.
- Press [(] [6] [−] [8] [)] [x²] [(] [5] [−] [7] [)] [^] [3].
- Press [ENTER] to see the result, -32.

Evaluate $8-3[-2(5-7)-5(4-2)]$.

- Note that only parentheses may in used in computations on the calculator.
- Press [8] [−] [3] [(] [(-)] [2] [(] [5] [−] [7] [)] [−] [5] [(] [4] [−] [2] [)] [)] [)].
- Press [ENTER] to see the result, 26.
-

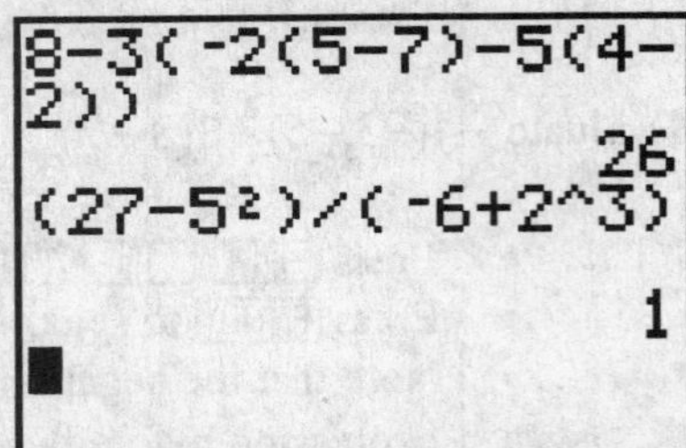

Evaluate $\frac{27-5^2}{-6+2^3}$.

- When a fraction has more than one term in the numerator and/or the denominator, you must enclose the numerator in parentheses and the denominator in parentheses.
- Press [(] [2] [7] [−] [5] [x²] [)] [÷] [(] [(-)] [6] [+] [2] [^] [3] [)].
- Press [ENTER] to see the result, 1.

CHAPTER 3

EVALUATING ALGEBRAIC EXPRESSIONS

General Guidelines for Evaluating Algebraic Expressions

- Enter the expression in the same order that you would write it.
- Familiarize yourself with the operation keys on the keyboard.
- Familiarize yourself with the operations found in the Math menu, e.g. absolute value, converting a decimal to fraction, etc. Press the [MATH] key to view this menu.
- Use the [X,T,Θ,n] key to enter the variable x.
- Use parentheses when you have more than one term in a radical, absolute value, or argument of a function, and in the numerator or denominator of a fraction.
- Use (-) the negation key [(-)] when the first term of an expression is negative. The negation key is to the left of the [ENTER] key.
- Use the subtraction key [−] between terms.
- TI calculators understand implied multiplication such as $5(1+2)$.

```
-2(5+6)²+√(9)
                -239
(3-4²)/(6+2)
              -1.625
Ans►Frac
               -13/8
```

To Enter Algebraic Expressions

Here are some examples of algebraic expressions and how they are entered. More examples and details about how to find and perform these operations will be shown as needed.

- Algebraic expressions are entered in the same order that you write them.
- Special functions such as $\sqrt{4}$, $\ln 5^2$ are entered in the same order that you write them.
- Use [2nd] [x^2] [4] to enter $\sqrt{}(4)$ as shown.
- Press [ENTER] to see the result, 2.
- Use [LN] [5] [x^2] to enter $\ln 5^2$ as shown.
- Press [ENTER] to see the result, 3.218875825.

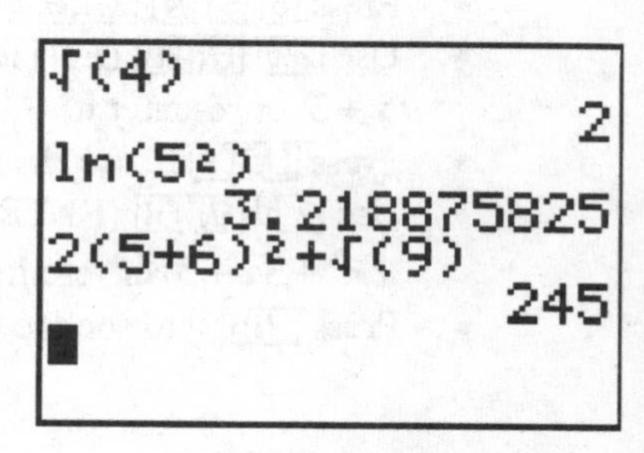

- Only parentheses can be used as grouping symbols of algebraic expressions.
- Brackets [ ] are reserved for matrices, and braces { } are reserved for lists. You will get an error message if you use brackets or braces in an algebraic expression.
- Enter $2(5+6)^2+\sqrt{9}$.
- Press [2] [(] [5] [+] [6] [)] [x^2] [+] [2nd] [x^2] [9].
- Press [ENTER] to see the result, 245.

Use the [STO▸] Key to Store a Value to a Variable on the Home Screen

The [STO▸] key can be used on the Home screen when you wish to substitute a value for a variable in an algebraic expression. You may store a value to the variable x using the [X,T,Θ,n] key, or you may store a value to another letter, say A or B, using the [ALPHA] key.

Store a value to x using [STO▸] and [X,T,Θ,n], then evaluate an expression with x

Store the number 2 to x.

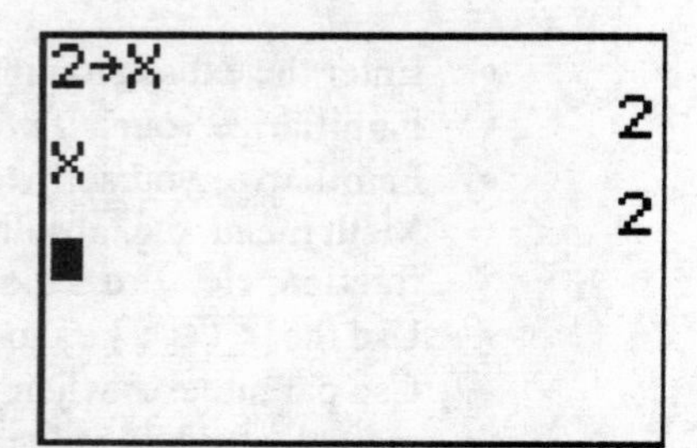

- First store 2 to x.
- Press [2] [STO▸] [X,T,Θ,n].
- You will see 2 →x on the Home screen.
- The number 2 will be used in place of x until you change it.
- Press [ENTER] to see just the number 2 on the next line.
- To verify that 2 is stored to x, press [X,T,Θ,n] [ENTER] to see the number 2 on the next line.

Evaluate $x+3$ and $2x^2+3x+5$ when $x=2$.

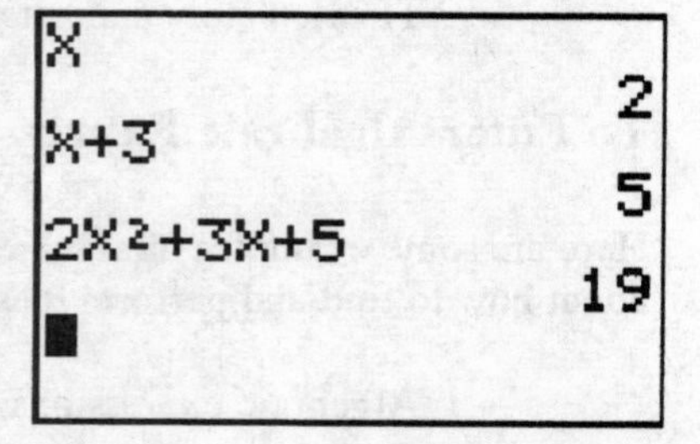

- Press [X,T,Θ,n] [ENTER] to verify that 2 is stored in x.
- Enter the expression $x+3$ on the Home Screen.
- Press [ENTER] to see the value, 5.
- Enter the expression $2x^2+3x+5$ on the Home screen.
- Press [ENTER] to see the value, 19.

Evaluate $x+3$ and $2x^2+3x+5$ when $x=6$.

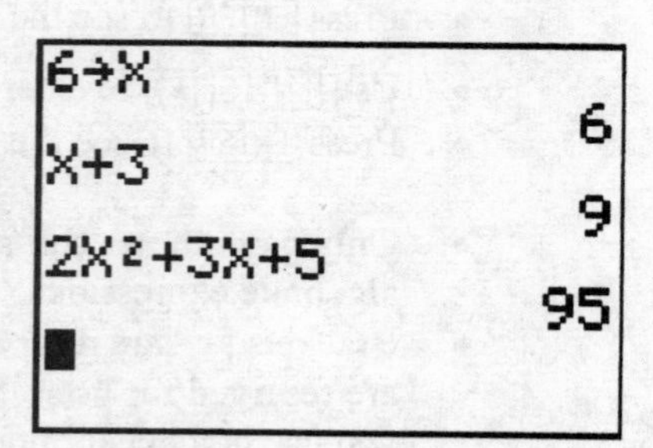

-
- Now store the number 6 to x and re-evaluate $x+3$ and $2x^2+3x+5$.
- Press [6] [STO▸] [X,T,Θ,n] [ENTER].
- Use [2nd] [ENTER] (ENTRY) until you find the expression $x+3$ or re-enter it.
- Press [ENTER] to see the new result, 9.
- Use [2nd] [ENTER] (ENTRY) until you find the expression $2x^2+3x+5$ or re-enter it.
- Press [ENTER] to see the new result, 95.

Store a value to another letter using [STO▶] [ALPHA], then evaluate an expression

You may store a number to a different letter, say A or B, by using the ALPHA key and the key that has the letter you want above it.

Store 2 to A, 3 to B, and 4 to C.

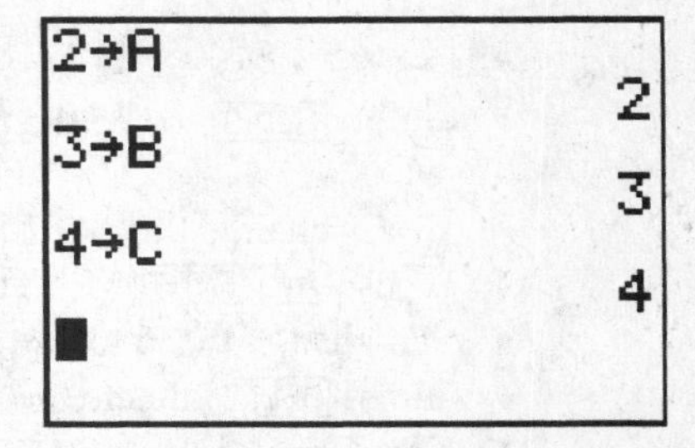

-
- To store 2 to A, press [2] [STO▶] [ALPHA] [MATH] (A).
- Press [ENTER] to see 2 on the next line.
- To store 3 to B, press [3] [STO▶] [ALPHA] APPS (B).
- Press [ENTER] to see 3 on the next line.
- To store 4 to C, press [4] [STO▶] [ALPHA] [PRGM] (C).
- Press [ENTER] to see 4 on the next line.

Evaluate $A+B+C$

- To enter $A+B+C$ on the Home Screen, press [ALPHA] [MATH] [+] [ALPHA] APPS [+] [ALPHA] [PRGM].
- Press [ENTER] to see the result, 9.

- You can also enter all the values to be stored on one line separated by the colon (:).
- Press the green [ALPHA] key, then the decimal point key [.] to enter a colon (:).
- Enter 2→A:3→B:4→C as shown.
- Enter the expression $A+B+C$.
- Press [ENTER] to see the result, 9.

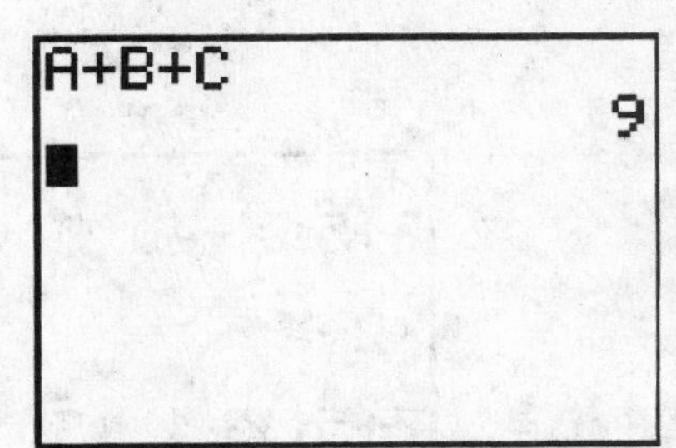

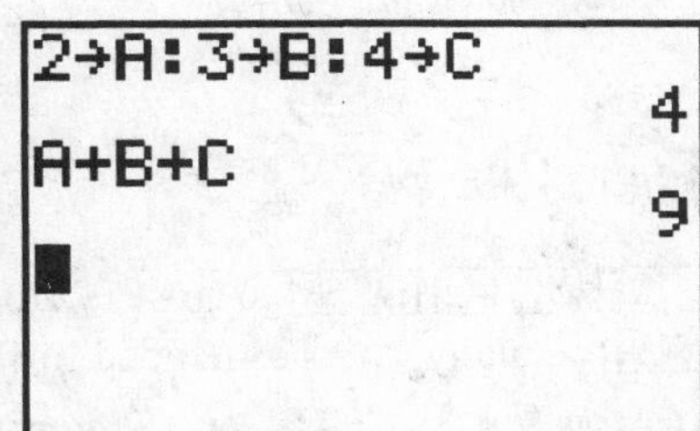

Evaluate an Expression at Different Values using [2nd] [ENTER] (ENTRY)

This is a quick way to evaluate an expression when you want to immediately do several computations using the same expression with different values.

- Enter $7+5(6-4)$^3 on the Home screen.
- Press [ENTER] to see the first result, 47.

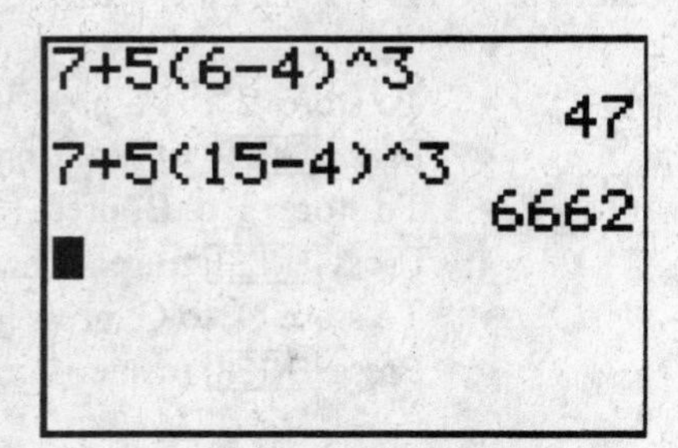

- Let's change the 6 to15.
- Press [2nd] [ENTER] (ENTRY) to repeat the expression.
- To change the 6 to 15, position the cursor over the 6.
- Press [DEL] to delete the 6.
- Next, use [2nd] [DEL] (INS) to insert 15.
- Press [ENTER] to see the new result, 6662.

Evaluate $7+5(x-4)^3$ for $x=6$ and $x=-2$.

```
6→X
                6
7+5(X-4)^3
               47
```

To evaluate the expression at $x=6$:

- Press [CLEAR] once or twice to clear the screen.
- Press [6] [STO▸] [X,T,Θ,n] [ENTER] to store 6 to x.
- Evaluate $7+5(x-4)^3$ for $x=6$.
- Press [7] [+] [5] [(] [X,T,Θ,n] [−] [4] [)] [^] [3].
- Press [ENTER] to see the result, 47.

To evaluate the expression at $x=-2$:

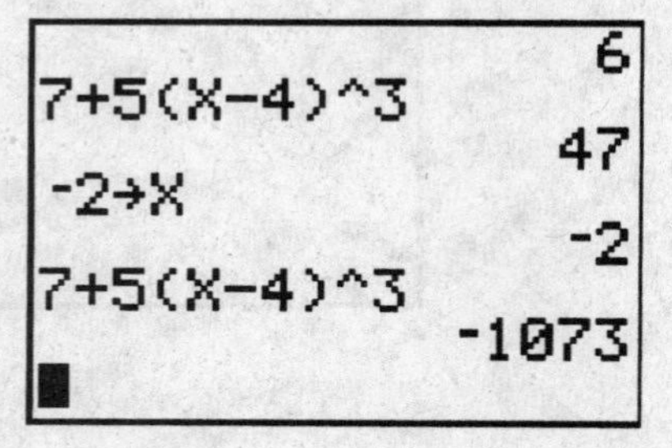

- Press [(-)] [2] [STO▸] [X,T,Θ,n] [ENTER] to store -2 to x.
- Press [2nd] [ENTER] as many times as needed to see the original expression $7+5(x-4)^3$, or re-enter it.
- Press [ENTER] to see the result, -1073.

Use the formula $C = \frac{5}{9}(F - 32)$ to convert the Fahrenheit temperature of $68^{\circ}F$ to its equivalent temperature on the Celsius scale.

- Press [CLEAR] once or twice to clear the screen.
- To store 68 to F, press [6] [8] [STO▸] [ALPHA] [COS].
- Press [ENTER].
- Press [(] [5] [÷] [9] [)] [(] [ALPHA] [COS] [−] [3] [2] [)] to enter the conversion formula $C = \frac{5}{9}(F - 32)$.
- Note that the fraction $\frac{5}{9}$ must be entered with parentheses as $(5/9)$.
- Note that [ALPHA] [COS] retrieves F.
- Press [ENTER] to see the result 20.
- This means that $68^{\circ}F$ is equivalent to $20^{\circ}C$.

Evaluate $x^2 - 5(x - y)$ for $x = 3$ and $y = 6$.

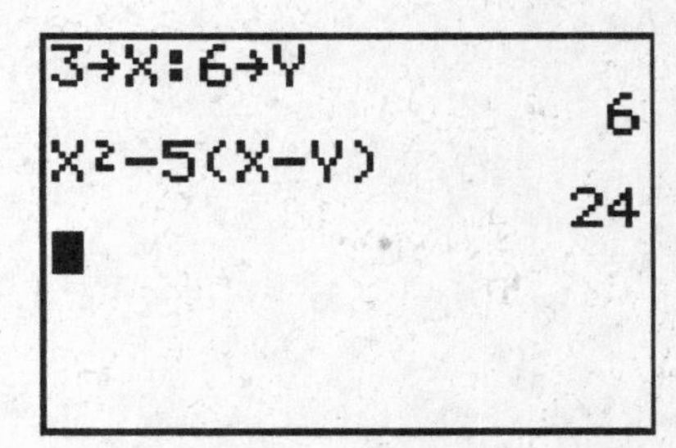

- Press [CLEAR] once or twice to clear the screen.
- Press [3] [STO▸] [X,T,Θ,n] [ALPHA] [.] [6] [STO▸] [ALPHA] [1] to store 3 to x and 6 to y on the same line.
- Press [ENTER].
- Press [X,T,Θ,n] [x²] [−] [5] [(] [X,T,Θ,n] [−] [ALPHA] [1] [)] to enter the expression $x^2 - 5(x - y)$.
- Press [ENTER] to see the result, 24.

CHAPTER 4

CHECK LINEAR EQUATIONS AND INEQUALITIES

CHECK A SOLUTION TO A LINEAR EQUATION

A value that makes an equation true is called a solution of the equation. Solutions to linear equations can be checked on the Home screen by replacing the value of the solution for x, or by using [STO▸]. Each example below is done both ways.

Check on the Home Screen by Replacement of x

Check that $x=9$ is the solution of $3x-1=26$.

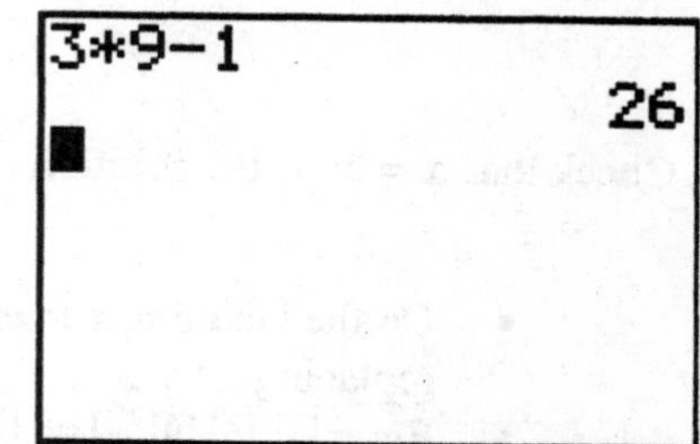

- On the Home screen enter the left side of the equation, replacing x by 9.
- Enter [3] [×] [9] [−] [1].
- Press [ENTER] to see the result, 26.
- 26 is the value of the right side of the equation.
- Hence $x=9$ is a solution of $3x-1=26$ because it makes the equation true.

Check that $x=2$ is the solution of $4(2x-3)+7=3x+5$.

```
4(2*2-3)+7
                11
3*2+5
                11
```

- On the Home screen enter the left side of the equation, replacing x by 2.
- Enter [4] [(] [2] [×] [2] [−] [3] [)] [+] [7].
- Press [ENTER] to see the result, 11.
- Enter the right side of the equation, replacing x by 2.
- Press [3] [×] [2] [+] [5].
- Press [ENTER] to see the result, 11.
- Hence $x=2$ is a solution of $4(2x-3)+7=3x+5$ because it makes the equation true.

Check that $x = -6$ is the solution of $\frac{3x}{2} = \frac{x}{5} - \frac{39}{5}$.

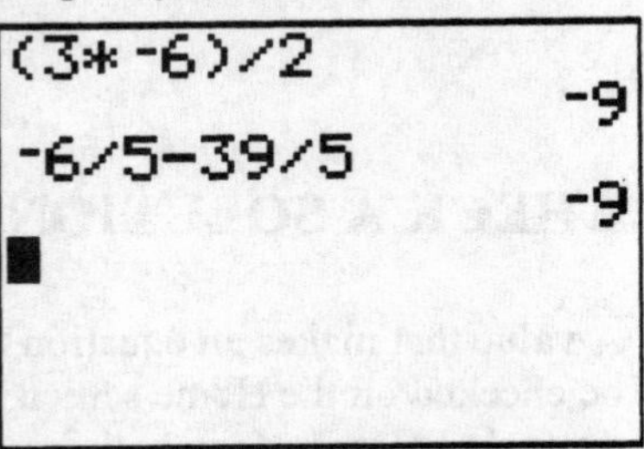

- **On the Home screen enter the left side of the equation, replacing x by −6.**
- **Enter [(] [3] [×] [(-)] [6] [)] [÷] [2].**
- **Press [ENTER] to see the result, −9.**

- **Enter the right side of the equation, replacing x by −6.**
- **Press [(-)] [6] [÷] [5] [−] [3] [9] [÷] [5].**
- **Press [ENTER] to see the result, −9.**

- **Hence $x = -6$ is a solution of $\frac{3x}{2} = \frac{x}{5} - \frac{39}{5}$ because it makes the equation true.**

Check that $x = 59$ is the solution of $\frac{x-2}{3} - 4 = \frac{x+1}{4}$.

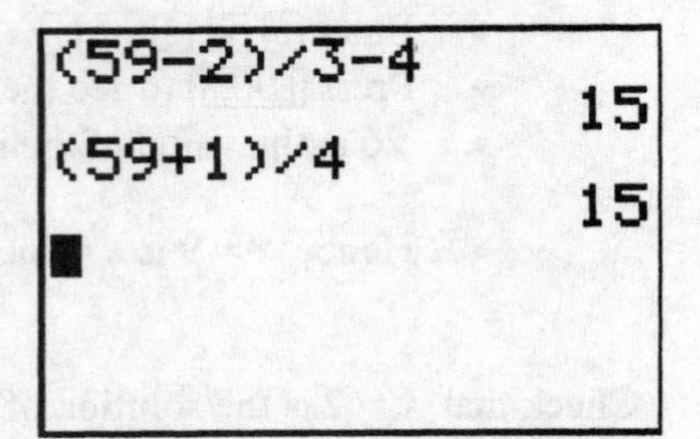

- **On the Home screen enter the left side of the equation, replacing x by 59.**
- **Enter [(] [5] [9] [−] [2] [)] [÷] [3] [−] [4]**
- **Press [ENTER] to see the result, 15.**

- **Enter the right side of the equation, replacing x by 59.**
- **Press [(] [5] [9] [+] [1] [)] [÷] [4].**
- **Press [ENTER] to see the result, 15.**

- **Hence $x = 59$ is a solution of $\frac{x-2}{3} - 4 = \frac{x+1}{4}$ because it makes the equation true.**

Check on the Home Screen Using [STO▸]

A value that makes an equation true is called a solution of the equation.

Check that $x = 9$ is the solution of $3x - 1 = 26$.

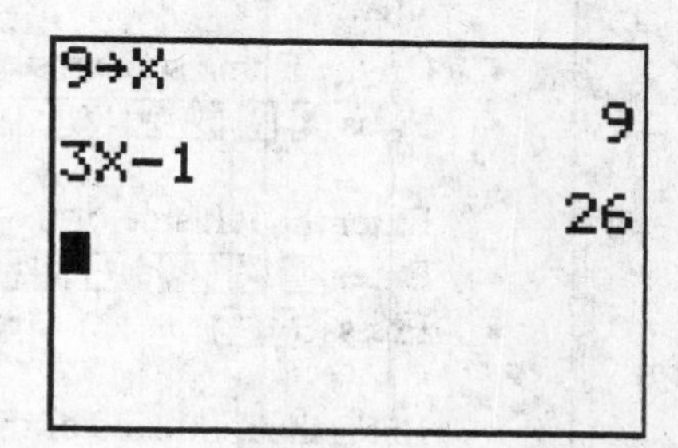

- On the Home screen, store 9 to x.
- Press [9] [STO▸] [X,T,Θ,n].
- Enter the left side of the equation.
- Press [3] [X,T,Θ,n] [−] [1].
- Press [ENTER] to see the result, 26.
- 26 is the value of the right side of the equation.
- Hence $x = 9$ is a solution of $3x - 1 = 26$ because it makes the equation true.

Check that $x = 2$ is the solution of $4(2x - 3) + 7 = 3x + 5$.

- On the Home screen, store 2 to x.
- Press [2] [STO▸] [X,T,Θ,n].

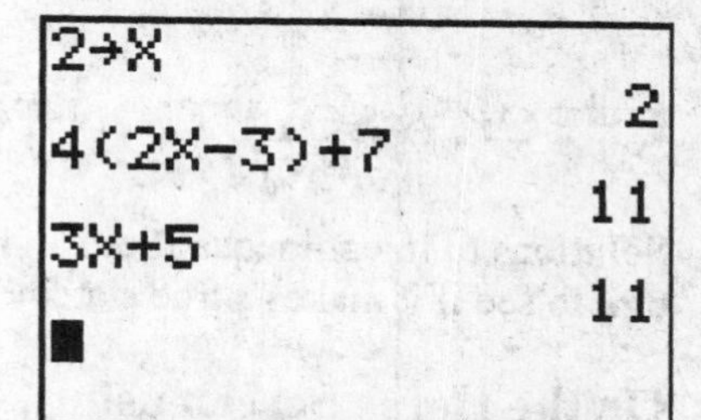

- Enter the left side of the equation.
- Enter [4] [(] [2] [X,T,Θ,n] [−] [3] [)] [+] [7].
- Press [ENTER] to see the result, 11.

- Enter the right side of the equation.
- Press [3] [X,T,Θ,n] [+] [5].
- Press [ENTER] to see the result, 11.

- Hence $x = 2$ is a solution of $4(2x - 3) + 7 = 3x + 5$ because it makes the equation true.

Check that $x = -6$ is the solution of $\frac{3x}{2} = \frac{x}{5} - \frac{39}{5}$.

- On the Home screen, store −6 to x.
- Press [(-)] [6] [STO▸] [X,T,Θ,n].

- Enter the left side of the equation.
- Enter [(] [3] [X,T,Θ,n] [)] [÷] [2].
- Press [ENTER] to see the result, −9.

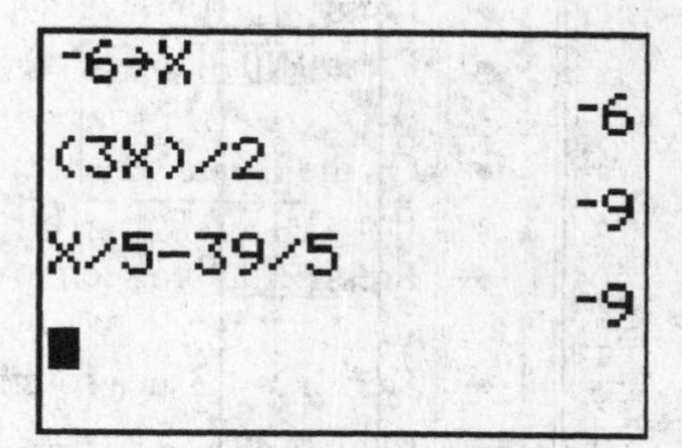

- Enter the right side of the equation.
- Press [X,T,Θ,n] [÷] [5] [−] [3] [9] [÷] [5]
- Press [ENTER] to see the result, −9.

- Hence $x = -6$ is a solution of $\frac{3x}{2} = \frac{x}{5} - \frac{39}{5}$ because it makes the equation true.

Check that $x = 59$ is the solution of $\frac{x-2}{3} - 4 = \frac{x+1}{4}$.

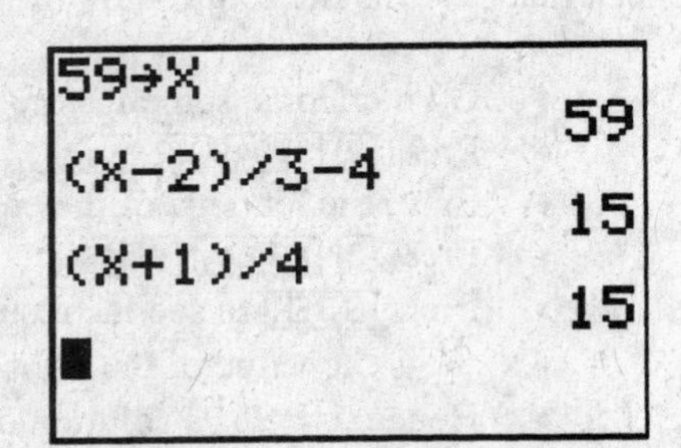

- On the Home screen, store 59 to x.
- Press [5] [9] [STO▸] [X,T,Θ,n].

- Enter the left side of the equation.
- Enter [(] [X,T,Θ,n] [−] [2] [)] [÷] [3] [−] [4]
- Press [ENTER] to see the result, 15.

- Enter the right side of the equation.
- Press [(] [X,T,Θ,n] [+] [1] [)] [÷] [4].
- Press [ENTER] to see the result, 15.
- Hence $x = 59$ is a solution of $\frac{x-2}{3} - 4 = \frac{x+1}{4}$ because it makes the equation true.

CHECK A SOLUTION TO A LINEAR INEQUALITY USING [STO▸]

Solutions to linear inequalities can be checked on the Home screen by using [STO▸] to replace a few values of x to see if it makes a true statement.

On the Home Screen using [STO▸]

Check that $x \geq 3$ is the solution of $4x - 7 \geq 5$.

First verify that $x = 3$ satisfies the inequality $4x - 7 \geq 5$.
Show that $x = 4$, a number to the right of 3, satisfies the inequality $4x - 7 \geq 5$.
Show that $x = 2$, a number to the left of $x = 3$, does not satisfy the inequality $4x - 7 \geq 5$.

First verify that $x = 3$ satisfies the inequality $4x - 7 \geq 5$.

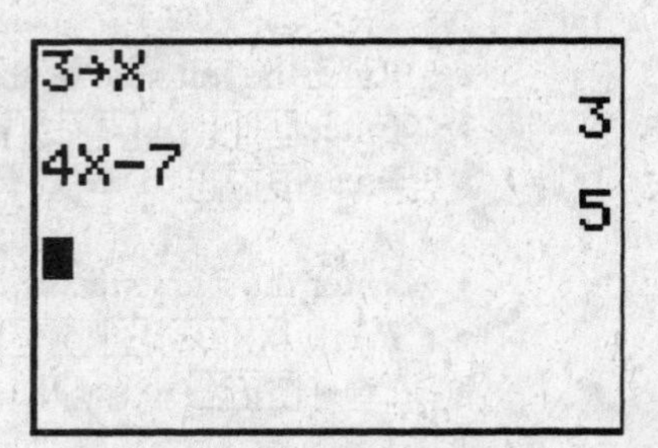

- On the Home screen store 3 to x.
- Press [3] [STO▸] [X,T,Θ,n].
- Press [ENTER] to see the value 3.

- On the Home screen enter the left side of the inequality.
- Press [4] [X,T,Θ,n] [−] [7].
- Press [ENTER] to see the value 5.

- Hence $x = 3$ is a solution of $4x - 7 \geq 5$ since $5 \leq 5$ is true because $5 = 5$.

Verify that $x = 4$, a number to the right of 3, satisfies the inequality $4x - 7 \geq 5$.

- On the Home screen store 4 to x.
- Press [4] [STO▶] [X,T,Θ,n].
- Press [ENTER] to see the value 4.
- On the Home screen enter the left side of the equation, $4x - 7$.
- Press [ENTER] to see the value 9.

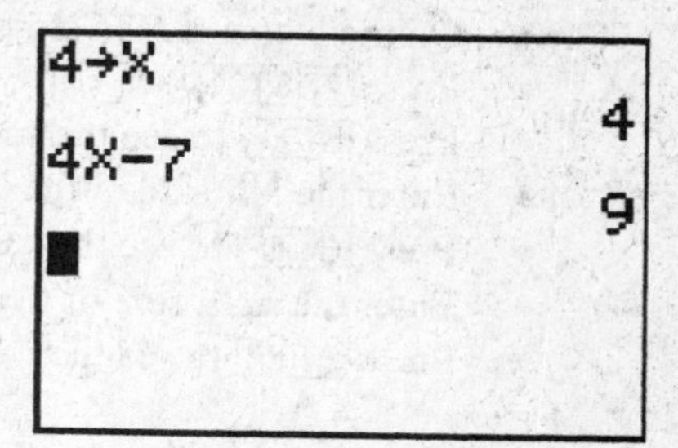

- Hence $x = 4$ is a solution of $4x - 7 \geq 5$ since $9 \geq 5$ is true.

Show that $x = 2$, a number to the left of $x = 3$, does not satisfy the inequality $4x - 7 \geq 5$.

- On the Home screen store 2 to x.
- Press [2] [STO▶] [X,T,Θ,n].
- Press [ENTER] to see the value 2.
- On the Home screen enter the left side of the equation, $4x - 7$.
- Press [ENTER] to see the value 1.

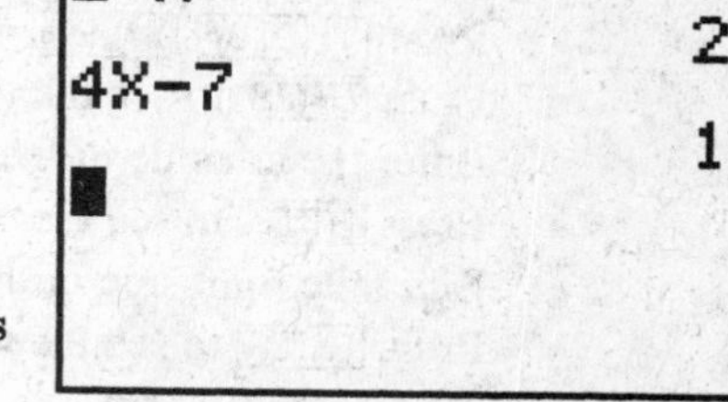

- Hence $x = 2$ is not a solution of $4x - 7 \geq 5$ since $1 \geq 5$ is not true.

- We conclude that $x \geq 3$ is the solution of $4x - 7 \geq 5$.

Check that $x > 2$ is the solution of $2x - 5 < 5x - 11$.

First find the value of the left side of the inequality, $2x - 5$ for $x = 3$.

- On the Home screen store 3 to x.
- Press [3] [STO▶] [X,T,Θ,n].
- Press [ENTER] to see the value 3.
- On the Home screen enter the left side of the inequality, $2x - 5$.
- Press [ENTER] to see the value 1.

Next, find the value of the right side of the inequality $5x - 11$ for $x = 3$.

- The value 3 is already stored to x.
- Enter the right side of the inequality, $5x - 11$.
- Press [ENTER] to see the value 4.

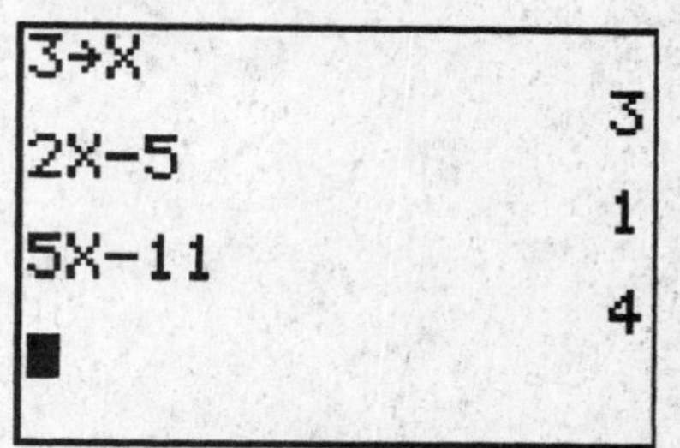

Compare the two results to see if the inequality is true.

- The inequality $2x - 5 < 5x - 11$ is true for $x > 2$ because $1 < 4$ is true.

Chapter 4: Check Linear Equations and Inequalities

Show that $x = 2$ does not satisfy the inequality $2x - 5 < 5x - 11$.

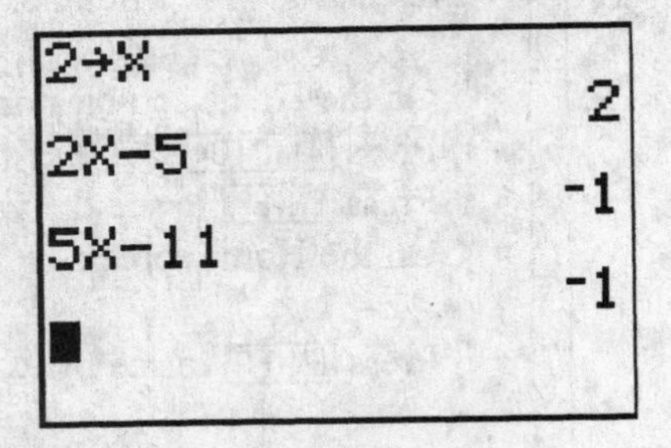

- On the Home screen store 2 to x.
- Press [2] [STO▸] [X,T,Θ,n].
- Press [ENTER] to see the value 2.
- Enter the left side of the inequality, $2x - 5$.
- Press [ENTER] to see the value -1.
- Enter the right side of the inequality, $5x - 11$.
- Press [ENTER] to see the value -1.

- Note that this is a not a true statement, since $-1 < -1$ is not true.

Show that $x = 1$, a number to the left of $x = 2$, such as does not satisfy the inequality $2x - 5 < 5x - 11$.

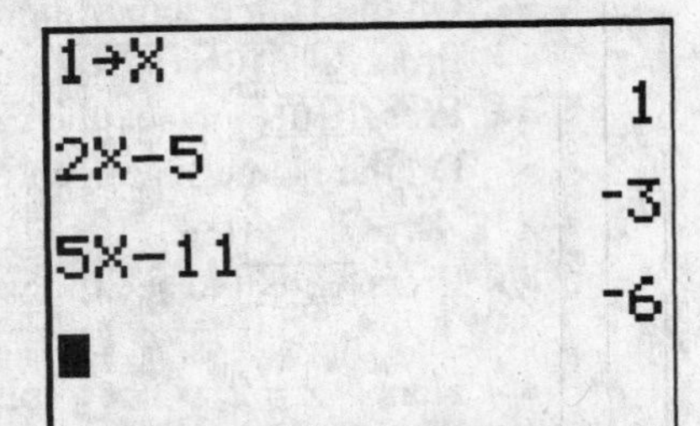

- On the Home screen store 1 to x.
- Press [1] [STO▸] [X,T,Θ,n].
- Press [ENTER] to see the value 1.
- Enter the left side of the inequality, $2x - 5$.
- Press [ENTER] to see the value -3.
- Enter the right side of the inequality, $5x - 11$.
- Press [ENTER] to see the value -6.

- Note that this is a not a true statement, since $-3 < -6$ is not true.

- We conclude that the inequality $2x - 5 < 5x - 11$ is true for $x > 2$.

CHAPTER 5

INTRODUCTION TO GRAPHING

GRAPHING

Graphing Keys

Graphing on the TI-83 Plus and the TI-84 Plus is handled in the same way on both calculators. **There are** five keys just below the screen: [Y=], [WINDOW], [ZOOM], [TRACE], and [GRAPH].

- You may input up to 10 functions in the [Y=] menu. Each function is identified by number, Y1=, Y2=, …, Y9=, Y0=.
- You may also graph a set of data points using Plot 1, Plot 2, or Plot 3 found on the graph screen above the Y= functions.
- [WINDOW] sets the size of the viewing screen.
- [ZOOM] is a menu that allows you to zoom on the graph.
- [TRACE] allows you to "walk" along the graph.
- [GRAPH] is used to see the graph of a function.
- The [2nd] functions above the graphing keys allow you to turn plots on and off using STATPLOT, set up a table using TBLSET, format the graph screen using FORMAT, find and calculate important points on the graph using CALC, and see a table of values for active functions using TABLE.
- Each of these functions will be explained as needed in this manual.

Preliminaries to graphing

- Press the [MODE] key.
- Use the arrow keys to position the cursor over FUNC on the fourth line. **This setting will** graph Y= functions on the graph screen.
- Press [ENTER].
- The other settings on that line are used to graph parametric equations using PAR, polar equations using POL, or sequences using SEQ.

- The function *must* be solved for *y*.
- Use only the variable *x* in the equation.
- Input *x* with the [X,T,Θ,*n*] key.
- Use the subtraction key [−] between terms, for example, x^2-3x-5.
- Use the negation key [(-)] when the first term is negative, for example, $-x+2$.

Use the [X,T,Θ,n] key to enter the variable x in the Y= menu.

- Press the [Y=] key. This brings you to the screen where you enter a function to be graphed.
- To enter $3x+2$, put the cursor next to Y1=.
- Press [3] [X,T,Θ,n] [+] [2] as shown.

The [Y=] Key

Input the function $y = x + 2$

```
Plot1 Plot2 Plot3
\Y1=X+2
\Y2=
\Y3=
\Y4=
\Y5=
\Y6=
\Y7=
```

- Press the [Y=] key below the screen.
- Position the cursor to the right of \Y1=.
- Press [CLEAR] to erase the old Y1.
- Press [X,T,Θ,n] [+] [2].
- The = sign is now highlighted to show that Y1 is ready to be graphed. We say that the function is now turned *on*.

Input the function $y = -x^2 - 2x + 1$.

```
Plot1 Plot2 Plot3
\Y1=X+2
\Y2=-X²-2X+1
\Y3=
\Y4=
\Y5=
\Y6=
\Y7=
```

- Press the [Y=] key below the screen.
- Position the cursor to the right of \Y2=.
- Press [(-)] [X,T,Θ,n] [x^2] [−] [2] [X,T,Θ,n] [+] [1].
- The = sign is now highlighted to show that Y2 is ready to be graphed. We say that the function is now turned *on*.

Turning functions on and off

- A function must be *on* to be graphed.
- A function is *on* when the equal sign is highlighted.
- To toggle between *on* and *off*, use the arrow keys to position the cursor over the equal sign.
- Press [ENTER]. The function will now be *on* if was *off*, and *off* if it was *on*.

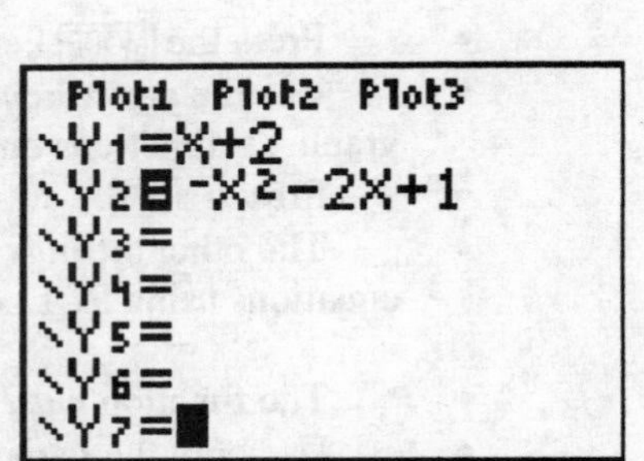

- Turn off Y1.
- To turn off Y1, place the cursor at Y1.
- Use the left arrow to place the cursor over the equal sign at Y1.
- Press [ENTER].
- The equal sign is now not highlighted. This means that Y1 will *not* be graphed.

Caution:

- If Plot 1, Plot 2, or Plot 3 is highlighted, you may get an error message or you may see more than the graph of your function, perhaps a histogram or a set of dots.
- Turn the Plot off by using the arrow keys to put the cursor over the Plot that is on.
- Press [ENTER] to turn the Plot off.

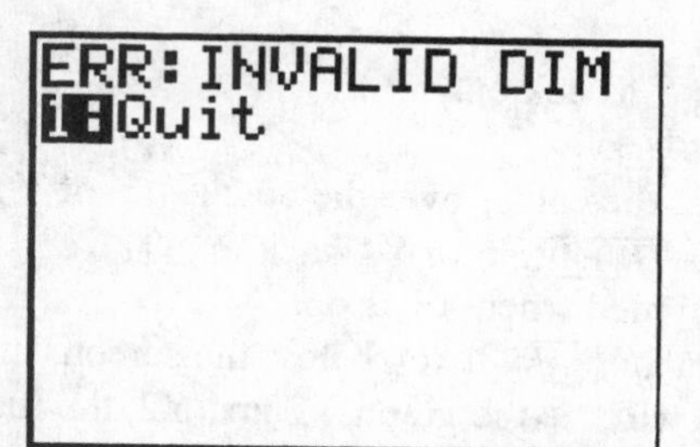

The [WINDOW] Key

Set the viewing window for Y2 = $-x^2 - 2x + 1$

```
WINDOW
 Xmin=-10
 Xmax=10
 Xscl=1
 Ymin=-20
 Ymax=5
 Yscl=2
 Xres=1
```

- The window setting defines the highest and lowest values of x and y on the graph of the function that will be shown on the screen. The window should show all the important parts of a graph.
- Press the [WINDOW] key below the screen.
- Use the arrow keys to position the cursor on the line you wish to change.
- Enter the values as shown. These settings are often referred to as [-10, 10, 1], [-20, 5, 2].
- The first set of numbers, [-10, 10, 1], means that the lowest x-value shown in the window is -10, called XMin, that the highest x-value shown in the window is 10, called XMax, and that each tick mark on the x-axis is 1 unit, called Xscl.
- Likewise, the second set of numbers, [-20, 5, 2] means that the lowest y-value shown in the window is -20, called YMin, that the highest y-value shown in the window is 5, called YMax, and that each tick mark on the y-axis is 2 units, called Yscl.
- Use Yscl = 2 so that there are not so many tick marks on the *y*-axis.

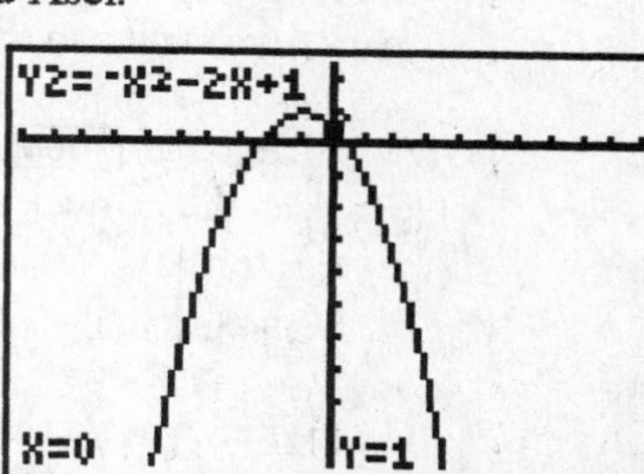

- Press [GRAPH] to see the graph of Y2 in this window.

The GRAPH key

Graph the functions Y1 and Y2 on the same screen

To graph both Y1 and Y2 on the same screen, both Y1 and Y2 must be turned on.

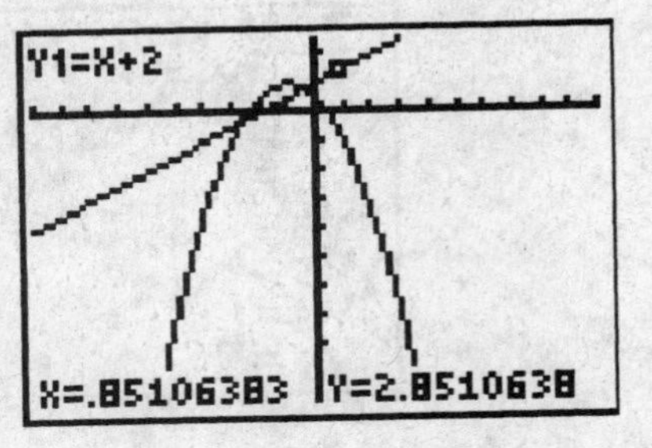

- Turn Y1 back on.
- Press Y=.
- Move the cursor over the equal sign of Y1.
- Press ENTER to turn Y1 back on. The equal sign will be highlighted when Y1 is on.
- Press the GRAPH key below the screen.
- You will see the graph Y1 and Y2, the functions that are turned *on*.

The TRACE Key

Trace the function

- Press the TRACE key below the screen.
- Use the left and right arrow keys to walk along the graph.
- The bottom of the screen displays the *x* and *y* coordinates of the point that the cursor is on.
- If more than one function is graphed on the same screen, use the up and down arrows to move from one function to another.

Display the equation on the graph

- The graph screen can be formatted by using the FORMAT feature found above the ZOOM key. In particular, you can display the equation.

- Press 2nd ZOOM (FORMAT).
- Use the arrow keys to place the cursor over the last line, ExprOn.
- Press ENTER.

- Whenever you use TRACE on a graph, the equation will be displayed.

The [ZOOM] Key and Special Windows

The Standard Window [-10, 10, 1] by [-10, 10, 1]

- The Standard window setting is often the first window that is used when graphing a function.
- The notation [-10, 10, 1] by [-10, 10, 1] means
  that the lowest x-value shown in the window is -10, called XMin,
  that the highest x-value shown in the window is 10, called XMax,
  and that each tick mark on the x-axis is 1 unit, called Xscl.
- Likewise, the second set of numbers means
  that the lowest y-value shown in the window is -10, called YMin,
  that the highest y-value shown in the window is 10, called YMax,
  and that each tick mark on the y-axis is 1 unit, called Yscl.

- Many functions can be graphed in the standard window.
- You may enter these values by hand or use the Zoom Standard feature of your calculator as follows.
- Press the [ZOOM] key.
- Move the cursor to 6:Zstandard and press [ENTER], or press the [6] key to select this setting.

- The function(s) that are *on* are automatically redrawn in the standard window.

- Here are the two functions graphed in the standard window.
- Note that the x and y values have several decimal places.

```
ZOOM MEMORY
1:ZBox
2:Zoom In
3:Zoom Out
4:ZDecimal
5:ZSquare
6:ZStandard
7↓ZTrig
```

```
WINDOW
 Xmin=-10
 Xmax=10
 Xscl=1
 Ymin=-10
 Ymax=10
 Yscl=1
 Xres=1
```

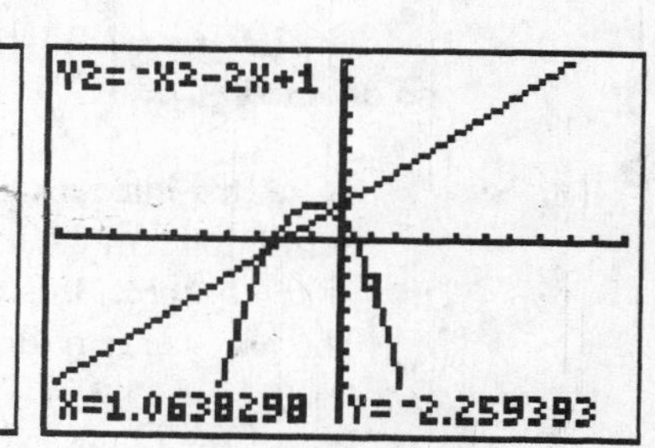

The Decimal Window [-4.7, 4.7, 1] by [-3.1, 3.1, 1]

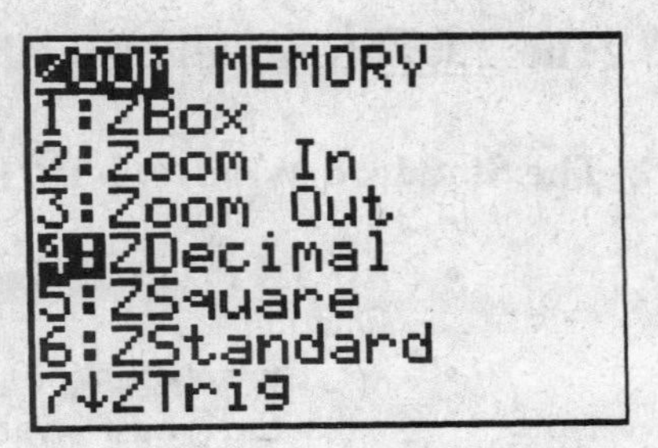

- The decimal window shows the x and y coordinates with one or two decimal places.

- To set the decimal window, press the [ZOOM] key.
- Scroll down to 4:Zdecimal and press [ENTER], or press the [4] key to select this window.
- The graphs are re-drawn in the new window.

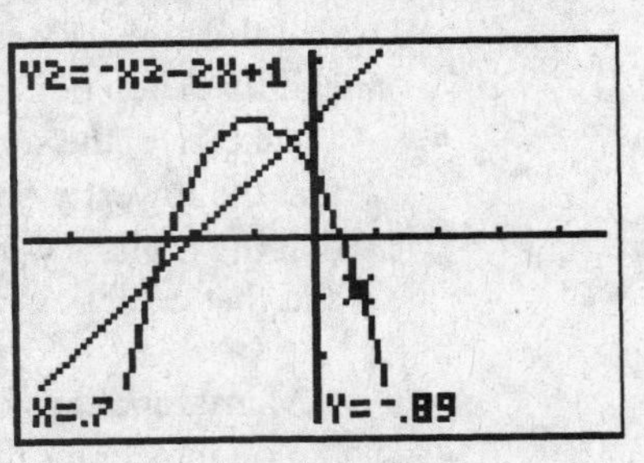

- Press [TRACE] to see how the coordinates are displayed.

- Use the left and right arrows to move around the function and observe that the coordinates are now shown with one or two decimal places.

- Use the up and down arrows to jump to the other function.
- Be aware that changing the window settings may take you out of the decimal window.

The Integer Window [-47, 47, 10] by [-31, 31, 10]

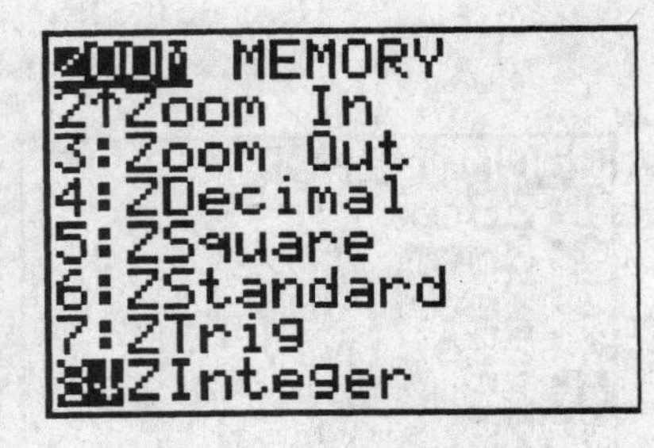

- The integer window shows the x and y values with no decimal places.

- To set the integer window, press the [ZOOM] key.
- Scroll down to 8:ZInteger and press [ENTER], or press the [8] key to select this window.
- The graphs are re-drawn in the new window.

- Press [TRACE] to see how the coordinates are displayed.
- Move around the curve and observe that the coordinates are now shown as integers.

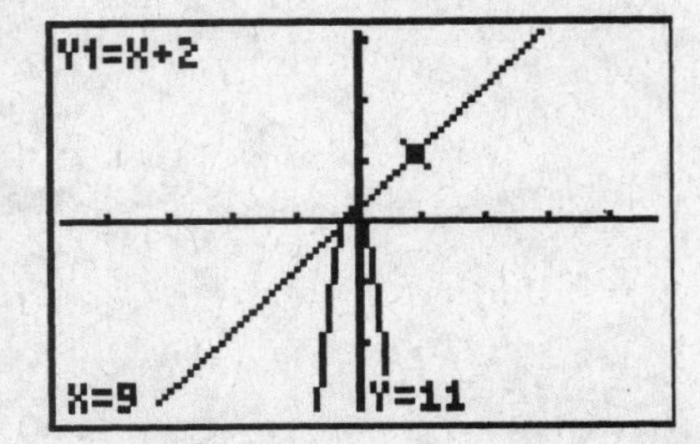

- Note that this is not the best window for displaying these two functions.
- Be aware that changing the window settings may take you out of the integer window.

CHAPTER 6

GRAPHING LINEAR EQUATIONS

GRAPH A LINEAR EQUATION

Graph $y = 2x + 3$ in the Standard Window

- Press the [Y=] key directly below the screen.
- Use the [CLEAR] key to clear Y1 if needed.
- Enter the $2x + 3$ as Y1.
- Press [GRAPH] to view the graph.

- Set the Standard window.
- Press [ZOOM] [6] to re-draw the graph in the standard window.
- Press [WINDOW] to see the window settings.

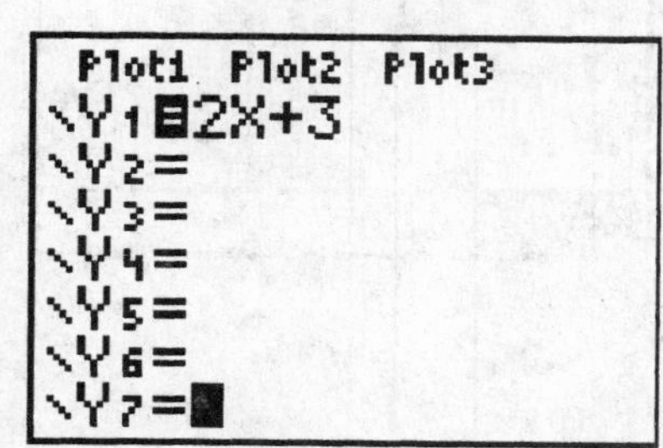

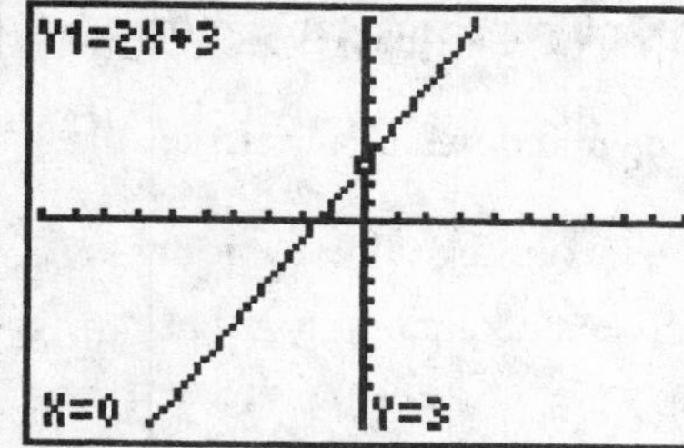

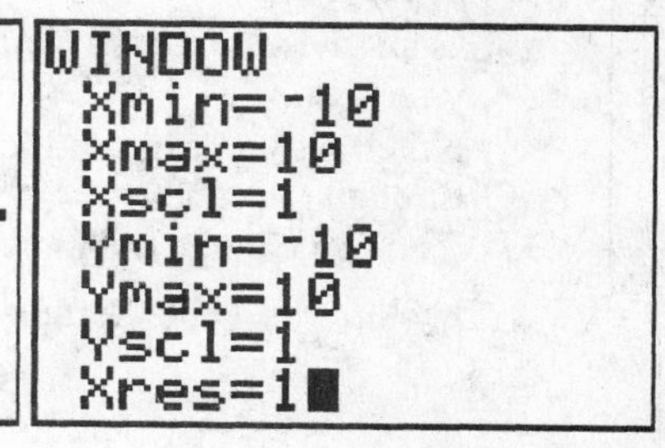

Display the equation on the graph

- The graph screen can be formatted by using the FORMAT feature found above the [ZOOM] key.
- In particular, you can display the equation on the graph screen.
- Press [2nd] [ZOOM] (FORMAT).
- Use the arrow keys to place the cursor over the last line, ExprOn.
- Press [ENTER].
- Now, whenever you use [TRACE] on a graph, the equation of the graph will be displayed.

Create an automatic table for the equation in Y1

- A table is created for each function that is turned on.
- Here, only the $Y1=2x+3$ should be turned on.
- Turn off or clear any other functions in the Y= menu.
- Press [2nd] [WINDOW] (TBLSET) to access the table setup. You can change any of these settings.
- Set TblStart= -3.
- Set △ Tbl= 1.
- Set **Indpnt: Auto** and **Depend: Auto** as shown. This will *automatically* give you table values for the Y= equations that are turned on for x-values starting at $x=-3$ with increments of 1.

TABLE SETUP
TblStart=-3
△Tbl=1
Indpnt: Auto Ask
Depend: Auto Ask

View the table

- Press [2nd] [GRAPH] (TABLE) to view the table.
- Use the up and down arrows to move along the x values of the table.
- Copy the values in the table on paper in a table form of your own.
- Note that you can use the up and down arrows to scroll through the table.
- You can change the increments of the table by going back to the Table Setup screen and changing △ Tbl.

X	Y1
-3	-3
-2	-1
-1	1
0	3
1	5
2	7
3	9

X= -3

Create a table for specific values of the equation in Y1

- Enter the function as $Y1=2x+3$.
- Set the table using [2nd] [WINDOW] (TBLSET) to access the table setup.
- Set TblStart= -3.
- Set △ Tbl= 1.
- Set **Indpnt: Ask and Depend: Auto** as shown. This will give you the y-values only for the x-values you give.
- Press [ENTER] to make the change from Auto to Ask.
- Press [2nd] [GRAPH] (TABLE) to view the table.
- Enter the x-values that you want to be evaluated in the table.
- Try the values $-4, 5, 10$ as shown.
- Press [ENTER] to see the corresponding y value.

X	Y1
-4	-5
5	13
10	23

X=

Find the x-intercept of a line

Approximate the x-intercept of $y = 2x + 3$ using [TRACE]

- The x-intercept is the point on the graph of the line that crosses the x-axis.
- The y-value of the x-intercept is always $y = 0$.
- Enter Y1 $= 2x + 3$.
- Press [GRAPH] to view the graph.
- Do you see the x-intercept of the line? If not, change the window until you do.
- Press [TRACE].
- Use the right arrow to get close to the x-intercept.
- Note that the cursor is "close" to the x-intercept, but not exactly on it.
- We see that $x \approx -1.5$ when $y \approx 0$.

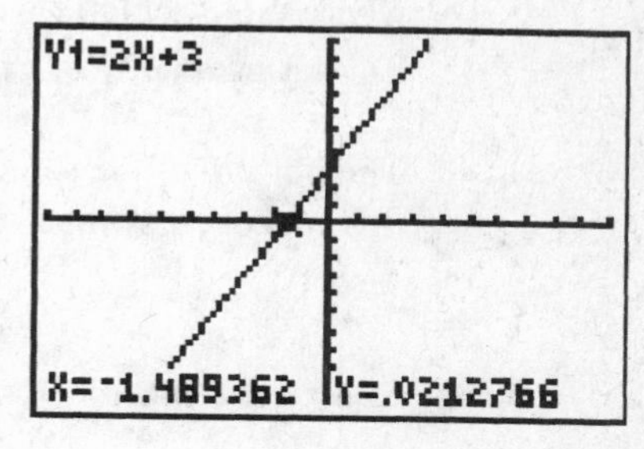

Find the exact coordinates of the x-intercept of $y = 2x + 3$ CALC menu

- Press [2nd] [TRACE] (CALC) to access the CALCULATE menu.
- Use the down arrow to scroll down to 2:zero and press [ENTER], or press the [2] key.

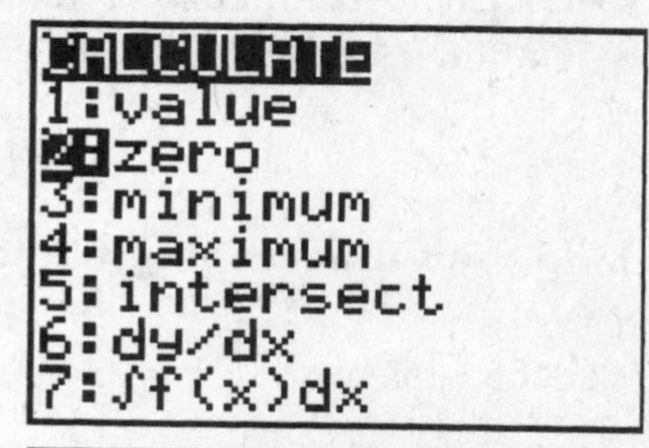

- The graph screen returns and asks for a Left bound?.
- The left bound defines the left side of the interval that includes the x-intercept. This question is asked because some functions have more than one x-intercept.
- Position the cursor at a point close to the x-intercept, but to the left of it.
- Press [ENTER].

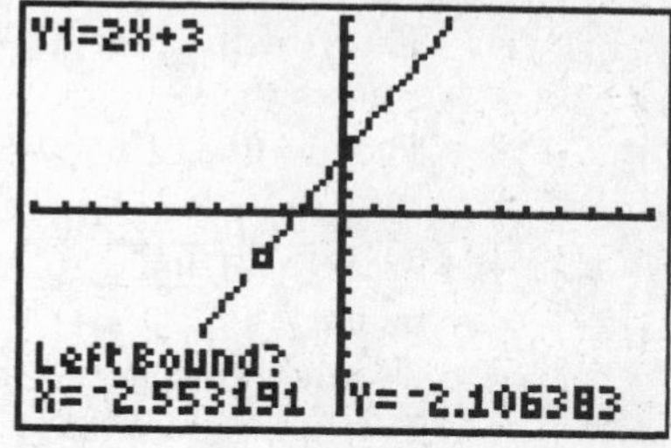

- Next, you see another screen that asks for the Right Bound?
- An arrow appears at the top of the screen.
- This arrow marks the left side of the interval that contains the x-intercept.

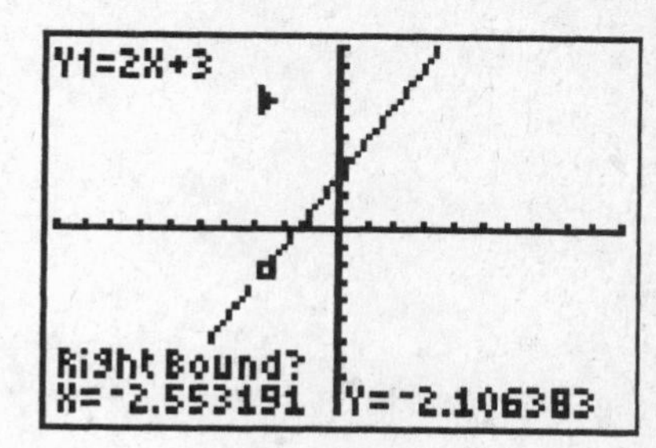

- Position the cursor close to the x-intercept, but to the right of it.
- You are defining the right side of the interval in which the x-intercept can be found.
- Press [ENTER].

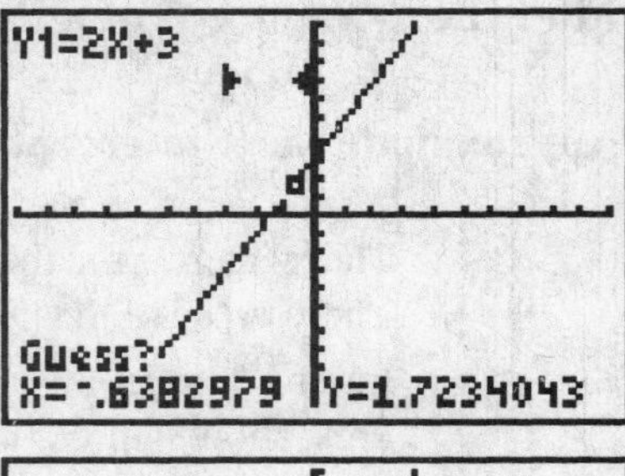

- The next screen asks for a guess of the x value.
- You can enter a value or just press [ENTER].
- The arrows that you see above the line define the interval where the calculator will look for an x-intercept.
- Enter a guess or just Press [ENTER].

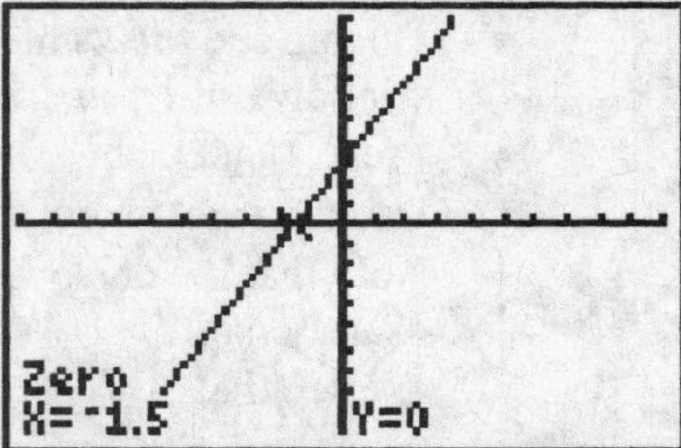

- The bottom of the screen shows that the x-intercept of this line occurs when $x = 1.5$ and $y = 0$, or at the point $(1.5, 0)$.
- Record the coordinates of the x-intercept.

***CAUTION:** Note that if you touch any other keys, you lose the result. Be sure to write down the result and a sketch of the graph before you do anything else on the calculator.*

Find the y-intercept of a line

Find the y-intercept of $y = 2x + 3$.

- The y-intercept is the point on the graph of the line that crosses the y-axis.
- The x-value of the y-intercept is always 0.

- Press [2nd] [TRACE] (CALC) to access the CALCULATE menu.
- Use the down arrow to scroll down to 1:value and press [ENTER], or press the [1] key.

- The graph screen returns and you see $x =$ flashing at the bottom of the screen.
- Press [0]. The equation will be evaluated when $x = 0$.
- Press [ENTER].
- The cursor jumps to the point $(0,3)$.
- The coordinates are displayed at the bottom of the screen.
- Record the y-intercept as the point $(0,3)$.

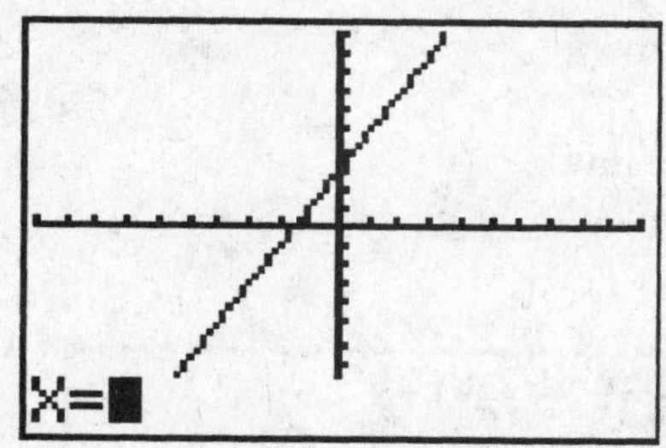

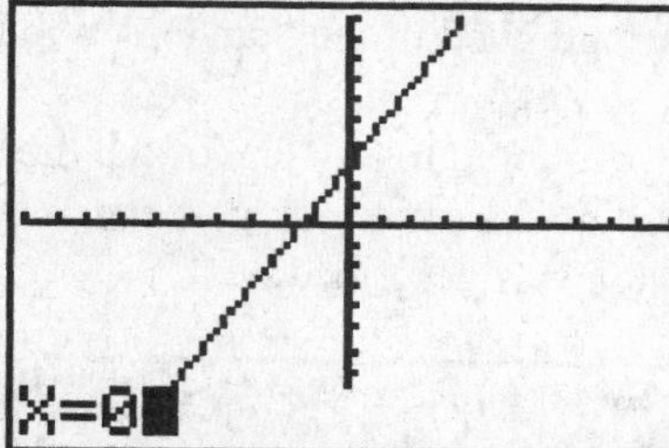

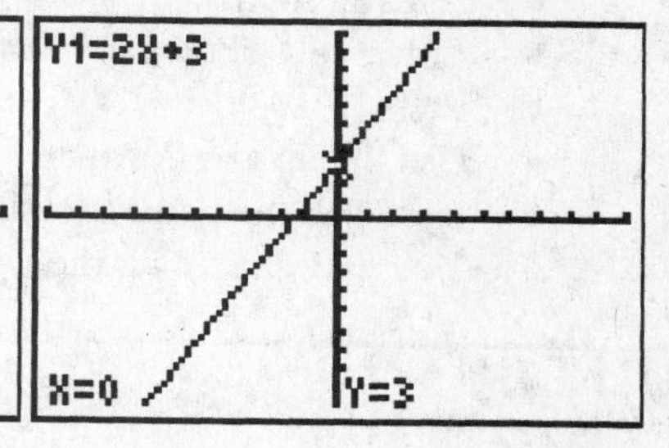

Graphing Parallel and Perpendicular Lines

Parallel and perpendicular lines are graphed by entering the equation of each line in the Y= menu. The graphs will look more accurate when the window is "Squared Up" using ZOOM SQUARE.

Graph the parallel lines $y = 2x + 1$ and $y = 2x + 8$ on the same screen

- Enter the equations to be graphed in the Y= menu.
- Press [Y=].
- Enter Y1 $= 2x + 1$.
- Enter Y2 $= 2x + 8$.
- Press [GRAPH].
- The graphs shown below are graphed in the Standard Window.
- Press [TRACE].
- Use the up and down arrows to toggle between the two lines.

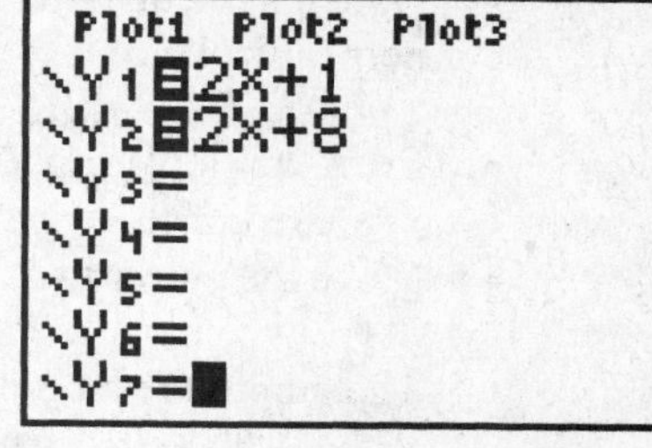

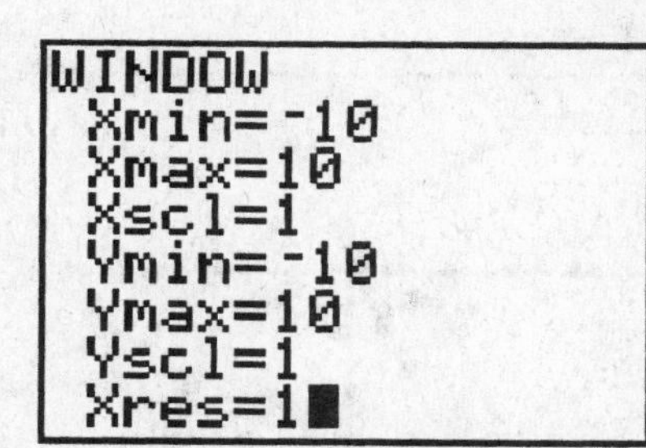

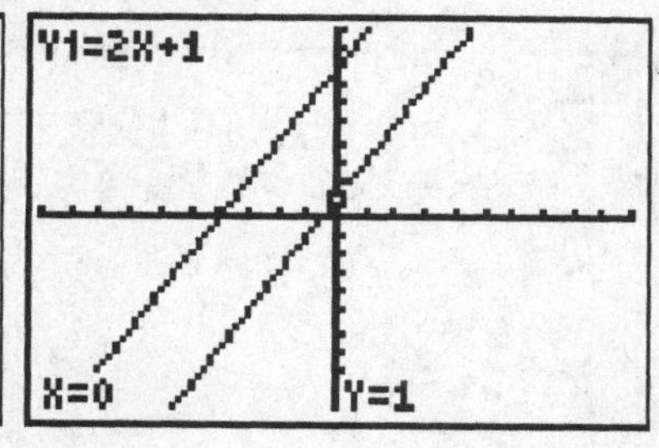

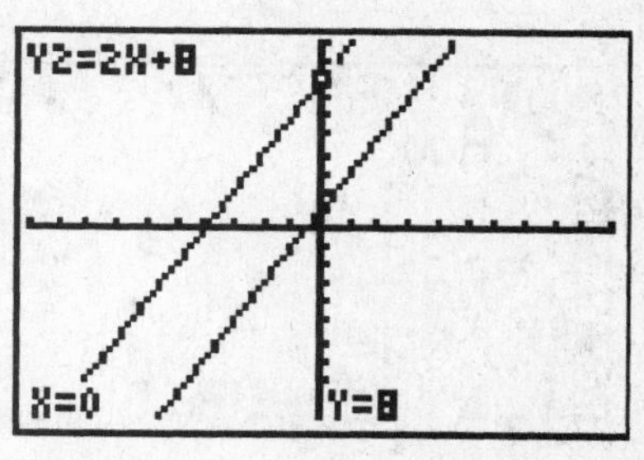

Graph $y = (-1/2)x + 3$, a line perpendicular to $y = 2x + 1$ and $y = 2x + 8$

- Press [Y=].
- Enter Y1 $= 2x + 1$.
- Enter Y2 $= 2x + 8$.
- Enter Y3 $= (-1/2)x + 3$.
- Press [GRAPH].
- The graphs shown below are graphed in the Standard Window.
- Press [TRACE].
- Use the up and down arrows to jump to any one of the three lines.
- Note that the perpendicular line does not look perpendicular.
- You may square up the window as described next.

Square up the window using Zoom Square

- Note that the parallel lines look reasonably parallel, but the perpendicular line does not look perpendicular to the other two lines.

- Press the [ZOOM] key.
- "Square up" the window by selecting 5:Zsquare in the ZOOM menu.
- Move the cursor to 5:Zsquare and press [ENTER], or press [5] to select Zoom Square.

- The graphs are automatically re-drawn in the new Zoom Square Window.
- Press [WINDOW] to view the settings of the Zoom Square Window.

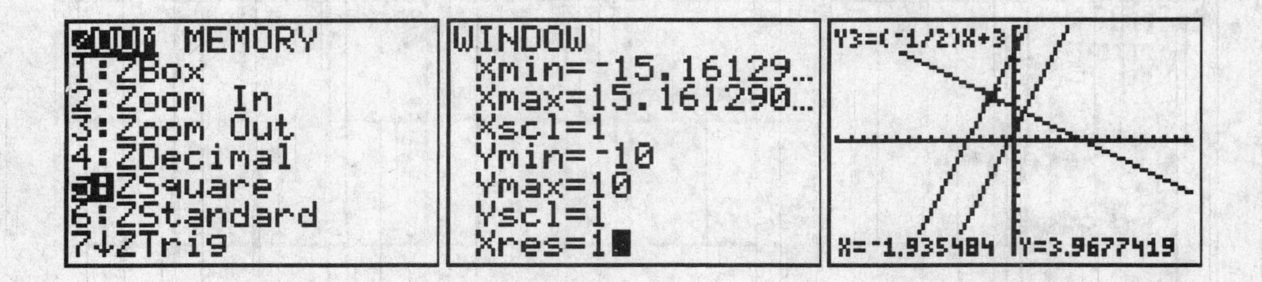

GRAPHING HORIZONTAL LINES

Graph a fixed Horizontal Line $y = b$ from the Y=screen

Graph $y = 2$

- Horizontal lines have equations that are solved for y, and so they may be graphed from the Y= screen.
- Press [Y=].
- Position the cursor at Y1.
- Press [CLEAR] the equation.
- Enter [2].
- Press [GRAPH] to view the graph of $y = 2$.
- Press [TRACE] and use the right and left arrows to move along the graph.

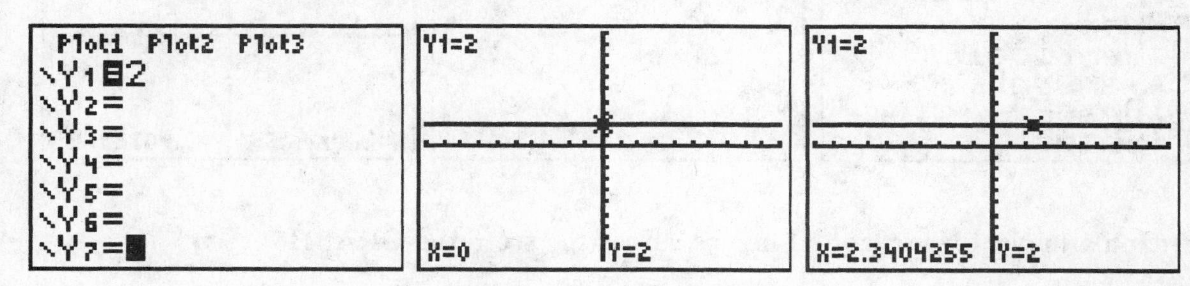

Graph a fixed horizontal line $y = b$ as a drawing from the Home screen

A Horizontal line may be drawn from the Home screen as a drawing rather than a function. You should clear any drawings before doing this.

Clear a Drawing using ClrDraw

- Before you attempt to draw a horizontal line from the Home screen, be sure that there are no drawings drawn on the graph screen.
- To clear all drawings, press [2nd] [PRGM] (DRAW) to access the DRAW menu.
- The cursor is on 1:C;rDraw.
- Press [ENTER] or [1] to put ClrDraw on the Home screen.
- Press [ENTER]. You will see the word Done.

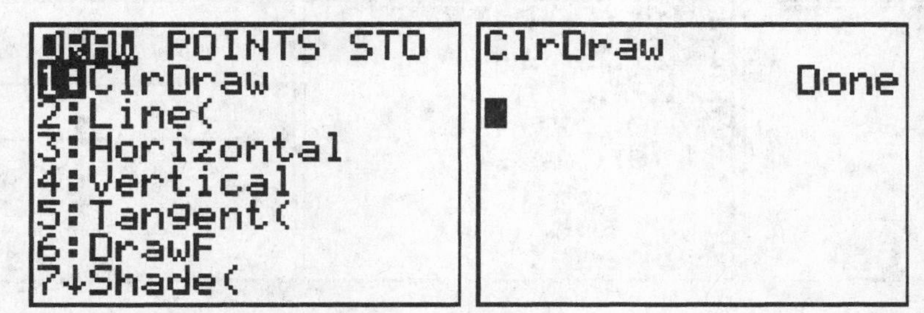

Graph a fixed horizontal line $y = b$ from the Home screen

- From the Home screen press [2nd] [PRGM] (DRAW) to access the Draw menu.
- In the DRAW submenu scroll down to 3:Horizontal or press the [3] key. This puts the word Horizontal on the Home screen.
- Press a number to draw the horizontal line. For example, the number 5 will graph the line $y = 5$.
- Press [ENTER] to view the graph.
- Note that [TRACE] does not work. You may use the arrow keys to move the cursor around the graph screen.

```
DRAW POINTS STO
1:ClrDraw
2:Line(
3:Horizontal
4:Vertical
5:Tangent(
6:DrawF
7↓Shade(
```

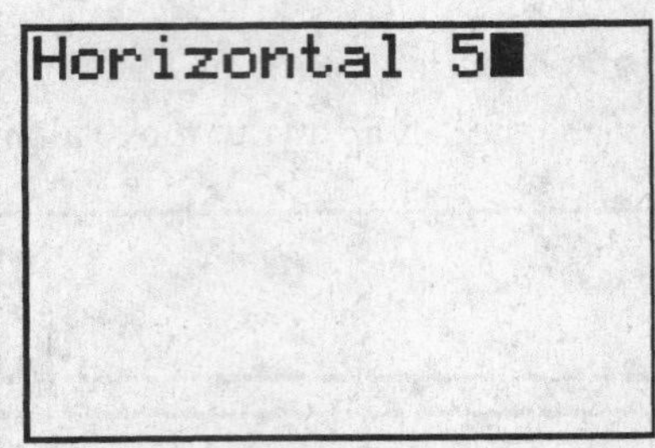

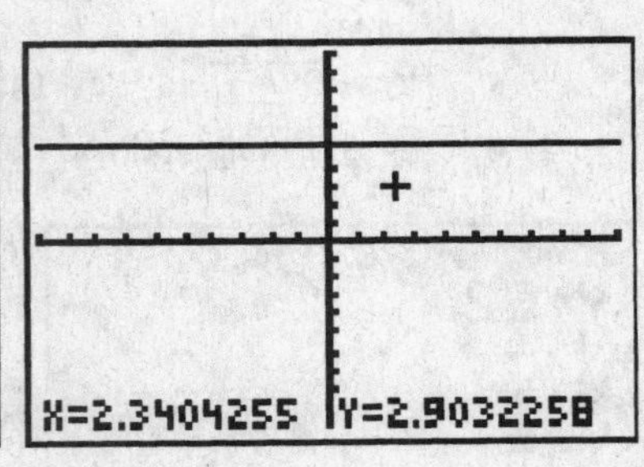

Graph a moveable horizontal line as a drawing from the GRAPH screen

- You may need to clear any drawings by selecting 1:ClrDraw in the DRAW menu.
- Press [ENTER] to see the word Done on the Home screen.
- Press [GRAPH] to get to the graph screen.
- From the **Graph** screen, **not the** Y= screen, press [2nd] [PRGM] (DRAW) to access the Draw menu.
- Select 3:Horizontal.
- Use the up and down arrow keys to move a horizontal line up and down.
- DO *NOT* PRESS [ENTER]. When you press [ENTER], the line is set in place, and is no longer movable. Note that [TRACE] does not work.
- Use the up and down arrow keys to move the horizontal line up and down.
- However, you can "fix" a movable horizontal by pressing [ENTER].

```
DRAW POINTS STO
1:ClrDraw
2:Line(
3:Horizontal
4:Vertical
5:Tangent(
6:DrawF
7↓Shade(
```

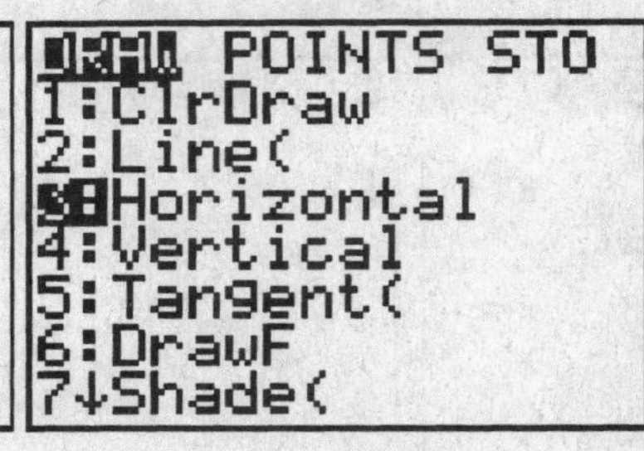

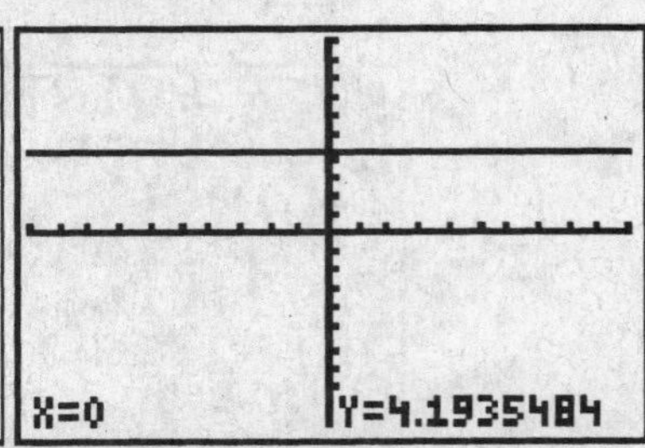

GRAPHING VERTICAL LINES

Graph a Vertical line $x = a$ as a drawing from the Home screen

Equations of vertical lines do not contain y , and cannot be graphed from the Y= screen. You may, however, show the graph of a vertical line as a drawing from the Home screen. Drawings cannot be traced and must be removed using ClrDraw as described below.

Clear a Drawing using ClrDraw

- Before you attempt to draw a vertical line, be sure that there are no drawings drawn on the graph screen.
- To clear all drawings, press [2nd] [PRGM] (DRAW) to access the DRAW menu.
- The cursor is on 1:C;rDraw.
- Press [ENTER] or [1] to put ClrDraw on the Home screen.
- Press [ENTER]. You will see the word Done.

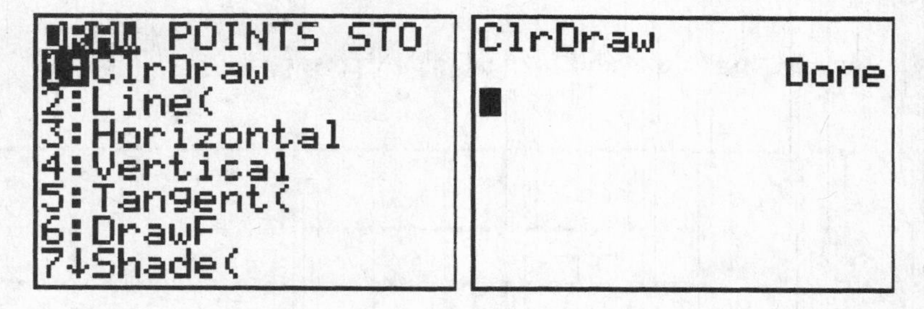

Graph a fixed vertical line $x = a$ from the Home screen

Graph $x = 5$

- A vertical line may be shown on the graph screen as a drawing.
- From the Home screen press [2nd] [PRGM] (DRAW) to access the Draw menu.
- In the DRAW submenu scroll down to 4:Vertical or press the [4] key. This puts the word Vertical on the Home screen.
- Press [5] next to the word Vertical.
- Press [ENTER] to view the graph.
- Note that [TRACE] does not work. You may use the arrow keys to move the cursor around the graph screen.
- Clear the drawing by selecting 1:ClrDraw in the DRAW menu.

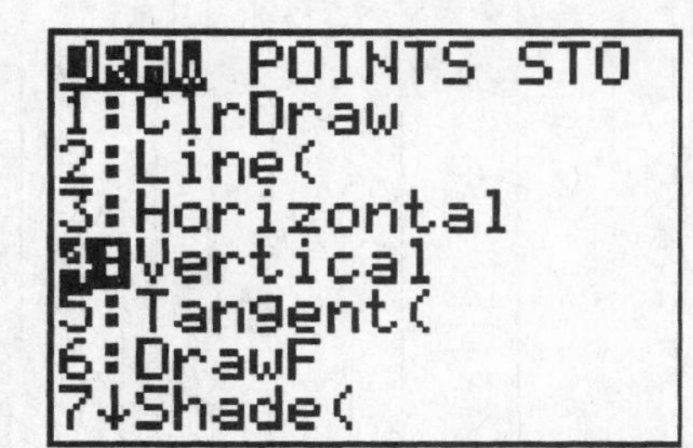

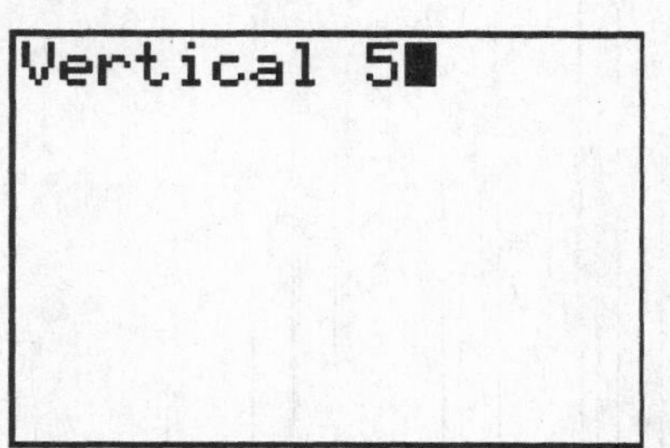

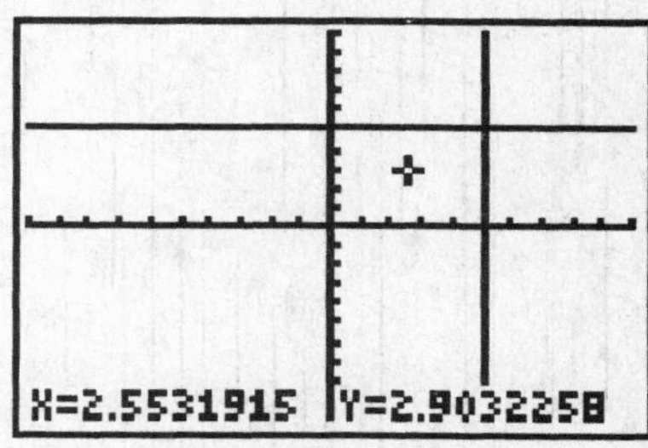

Graph a moveable vertical line from the GRAPH screen

- You may need to clear any drawings by selecting 1:ClrDraw in the DRAW menu.
- Press [ENTER] to see the word Done on the Home screen.
- Press [GRAPH] to get to the graph screen.

- From the **Graph** screen, **not the** Y= screen, press [2nd] [PRGM] (DRAW) to access the Draw menu.
- Select 4:Vertical.
- Use the left and right arrow keys to move the vertical line.

- DO *NOT* PRESS [ENTER].
- When you press [ENTER], the line is set and is no longer movable.
- Note that [TRACE] does not work.

- Note that you cannot have both a movable horizontal and movable vertical line on the graph screen at the same time.

- However, you can "fix" a movable vertical line by pressing [ENTER].

```
DRAW POINTS STO
1:ClrDraw
2:Line(
3:Horizontal
4:Vertical
5:Tangent(
6:DrawF
7↓Shade(
```

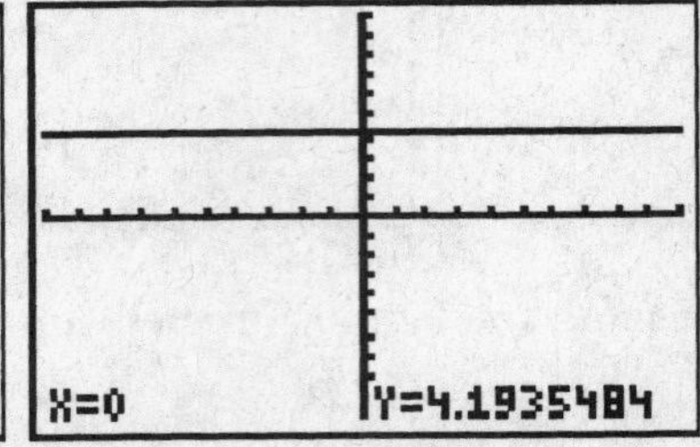

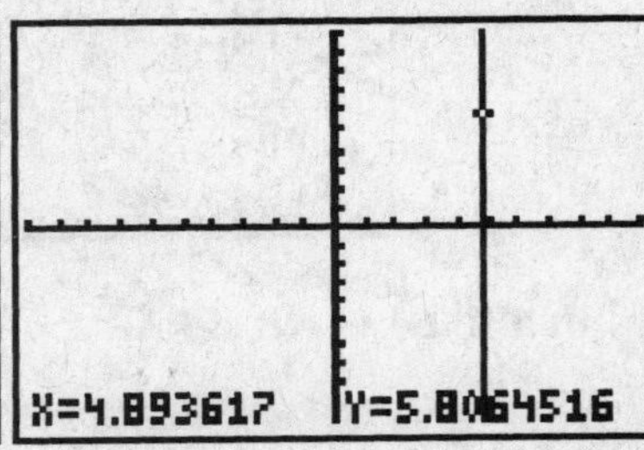

CHAPTER 7

GRAPHING FEATURES AND LINEAR EQUATIONS

SOLVE A LINEAR EQUATION WITH A TABLE

The problem: A plumber charges \$50 per house visit and \$35 per hour.

a) Write an equation for the plumber's total charge, y, given the number of hours x, for a single job.
b) Use a table to find the plumber's total charge for a job that could take 1 to 4 hours.

First write the equation

- The \$50 fee per house visit is represented by a constant value, 50.
- The plumber charges \$35 for each hour that he is at the house, so $35x$ represents the charge for the plumber's time.
- Therefore the equation that represents this situation is $y = 50 + 35x$.

Enter the equation as Y1

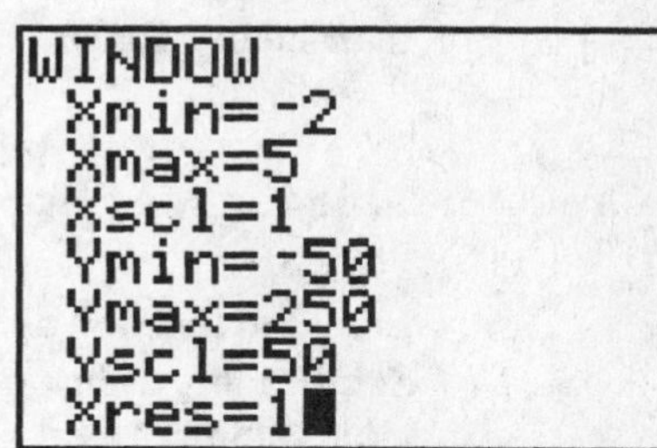

- Press the [Y=] key directly below the screen.
- Use the [CLEAR] key to clear Y1 if needed.
- Enter the $50 + 35x$ as Y1.
- Press [GRAPH] to view the graph.
- The y-values of the line must be visible in the window for x values between 1 and 4.
- If not, adjust the window.
- Here is a suggested window.
- The graph in this window is shown.

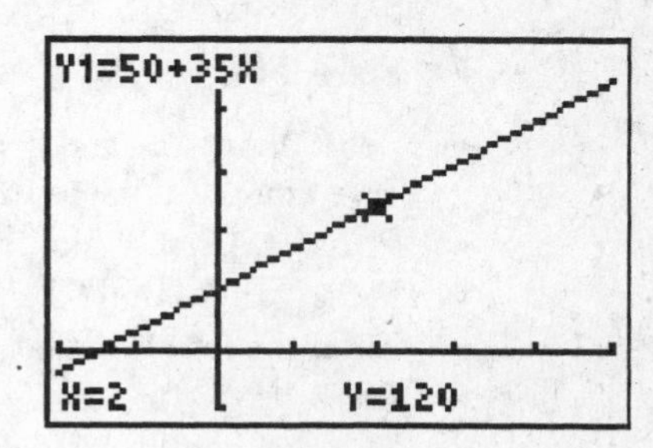

Create an automatic table for the equation in Y1

- A table is created for each function that is turned on.
- Here, only the $Y1 = 50 + 35x$ should be turned on.
- Turn off or clear any other functions in the Y= menu.

- Press [2nd] [WINDOW] (TBLSET) to access the table setup. You can change any of these settings.
- Set TblStart=1.
- Set Δ Tbl=1.
- Set Indpnt: Auto and Depend: Auto as shown. This will *automatically* give you table values for the Y= equations that are turned on for x-values starting at $x = 1$ with increments of 1.

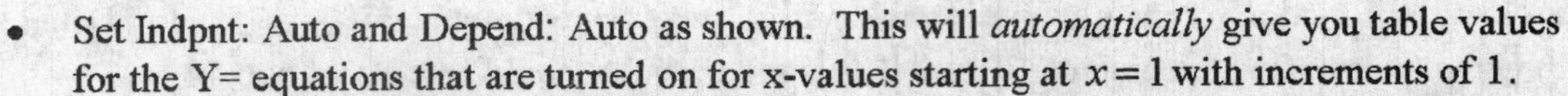

- Press [2nd] [GRAPH] (TABLE) to view the table.
- Use the up and down arrows to move along the x values of the table for $x = 1, 2, 3, 4$.
- Copy the values in the table on paper in a table form of your own.
- Note that you can use the up and down arrows to scroll through the table.
- You can change the increments of the table by going back to the Table Setup screen and change Δ Tbl.

- Hence, the plumber will charge $85 for 1 hour, $120 for 2 hours, $155 for 3 hours, and $190 for 4 hours. Record these values on your own paper.

Create a table for specific values of the equation in Y1

- Enter the function as $Y1 = 50 + 35x$.
- Set the table using [2nd] [WINDOW] (TBLSET) to access the table setup.
- Set TblStart=1.
- Set Δ Tbl=1.

- Set Indpnt: Ask and Depend: Auto as shown. This will give you the y-values only for the x-values you give.
- Press [ENTER] to make the change from Auto to Ask.
- Press [2nd] [GRAPH] (TABLE) to view the table.
- Enter the x-values that you want evaluated in the table.
- Press [ENTER] to see the corresponding y value.
- This is a good way to decide the window settings.

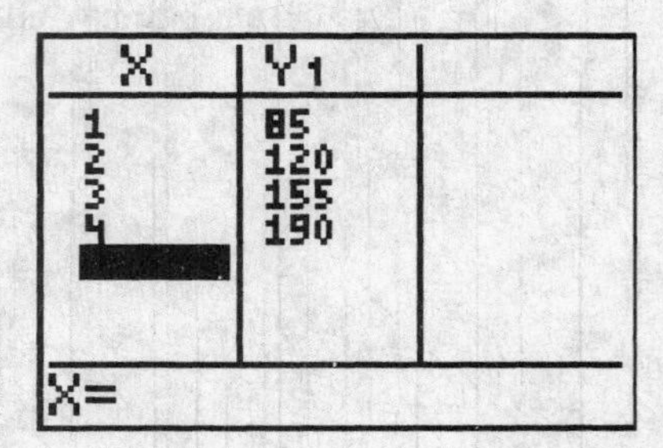

- Hence, the plumber will charge $85 for 1 hour, $120 for 2 hours, $155 for 3 hours, and $190 for 4 hours. Record these values on your own paper.

SOLVE A LINEAR EQUATION GRAPHICALLY

Solve a Linear Equation By Graphing Each Side of the Equation Separately

One way to solve an equation graphically is to consider each side of the equation as a separate function and find the intersection of the two functions.

Solve $3x+5=2x+13$.

First graph each side

- Press [ZOOM] [6] to set the standard window.
- Press the [Y=] key directly below the screen.
- Use the [CLEAR] key to clear Y1 and Y2 if needed.
- Enter the left side, $3x+5$ as Y1.
- Enter the right side, $2x+13$ as Y2.
- Press [GRAPH] to view the graphs.
- Do you see the intersection of the two lines?
- If not, adjust the window as described below. The graph shown on the right is graphed in the standard window.
- The cursor is on the graph of Y1, the left side of the equation.
- Use the up arrow [▲] key to move the cursor to the right side of the equation, the graph of Y2.

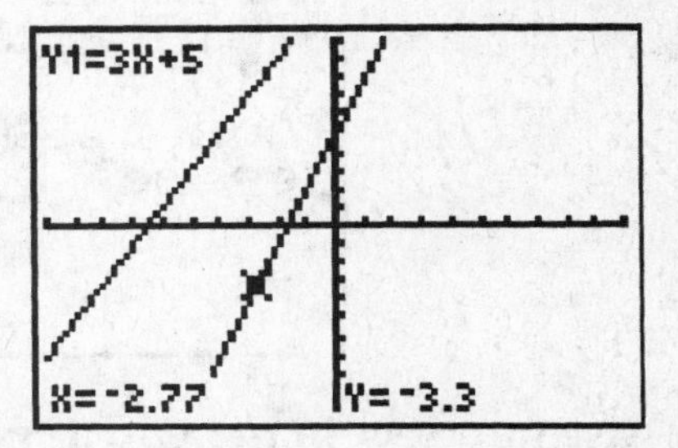

Find a window that shows the intersection point

- Press the [WINDOW] key to see the settings of x and y on the screen.
- The graphs shown above are graphed in the standard window. It looks like the intersection occurs somewhere to the right of $x=2$ and the y value is more than 10.
- Change the window settings until you see the intersection point, preferably near the middle of the screen.
- Here is a suggested window

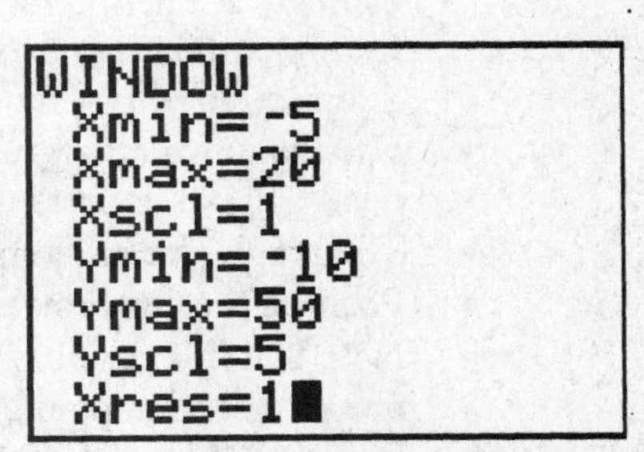

Find an approximate solution using TRACE

- Press [TRACE].
- Use the right arrow to get close to the intersection.
- Note that the cursor is not exactly on the intersection point. We see that $x \approx 7.5$ and $y \approx 28$.

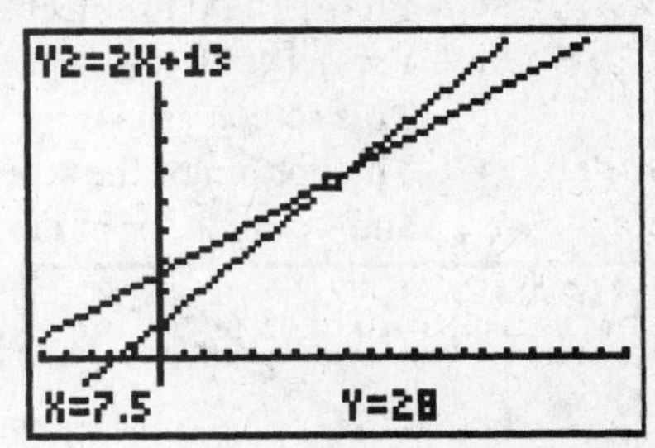

Use a table to locate the intersection point

- Set the table using [2nd] [WINDOW] (TBLSET). You can change any of these settings. For now, keep AUTO highlighted as shown. This will automatically give you table values for the Y= equations that are turned on.
- Press [2nd] [GRAPH] (TABLE) to view the table.
- Use the up and down arrows to move along the x values of the table, use the left and right arrows to move from Y1 to Y2.
- Notice that $x = 8$ has the same y value of 29 on both Y1 and Y2.
- So the intersection point, or the solution to the equation, is $x = 8$.
- Note that if the x-coordinate of the intersection point is not an integer, the table does not give you the intersection. You can look for it by changing Δ Tbl to a fractional value.

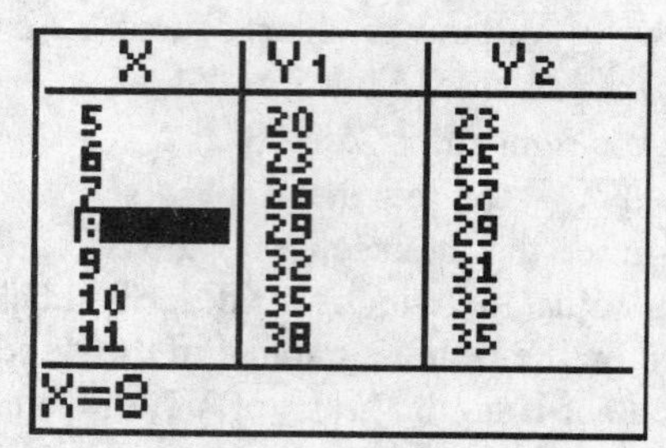

X	Y1	Y2
5	20	23
6	23	25
7	26	27
8	29	29
9	32	31
10	35	33
11	38	35

X=8

Use intersect in the CALC menu to find the exact coordinates of the intersection point

- Press [2nd] [TRACE] (CALC) to access the CALCULATE menu. This menu allows you to evaluate a function for a given value of x as well as to find important points on the graphs, such as the intersection of two functions.
- Use the down arrow to scroll down to 5:intersect and press [ENTER], or press the [5] key.
- The graph screen returns and asks for the first curve.
- Position the cursor near the intersection point.
- Note that it is possible for two curves to have more than one intersection point.
- Press [ENTER] to select the first curve as given.
- Press [ENTER] to select the second curve.
- This screen asks for a guess of the x value. You can enter a value or just press [ENTER].
- The bottom of the screen shows that the intersection of these two lines occurs when $x = 8$ and $y = 29$, or at the point $(8, 29)$.

CALCULATE
1:value
2:zero
3:minimum
4:maximum
5:intersect
6:dy/dx
7:∫f(x)dx

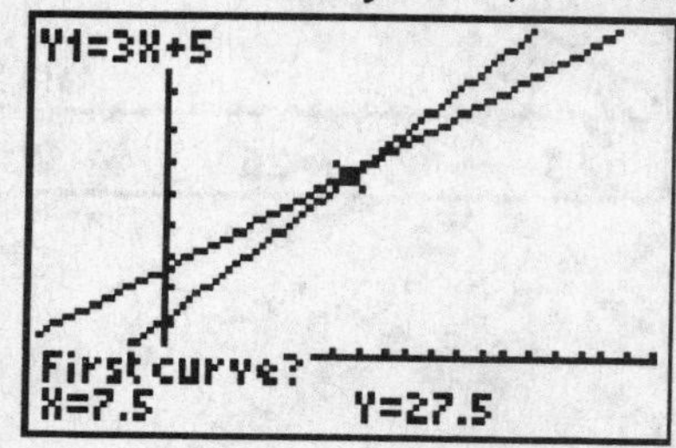

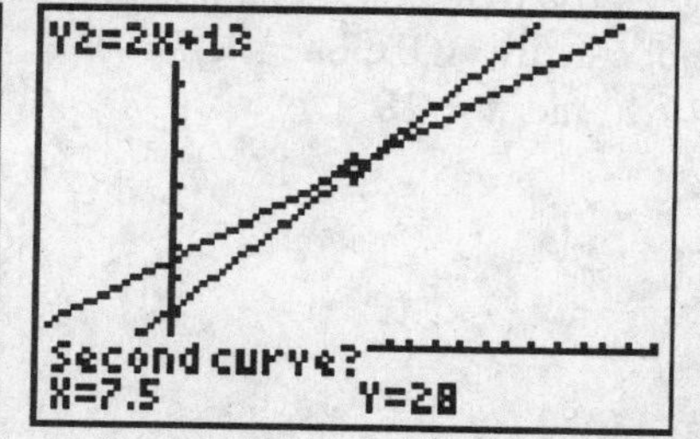

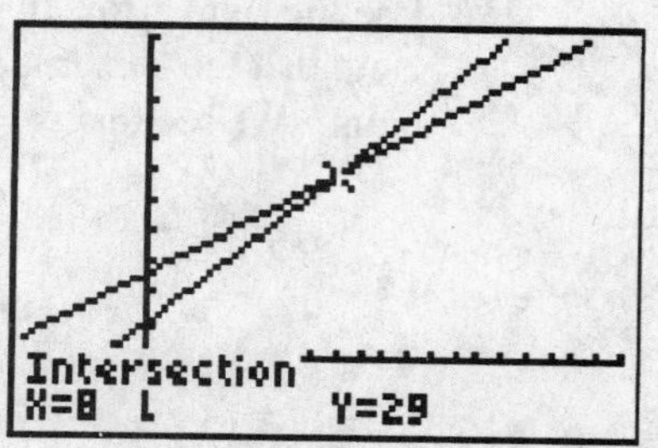

Solve a Linear Equation Using the x-intercept

When you solve an equation using the x-intercept, you must first collect all of the terms to one side of the equation, so that the other side is equal to 0.

Solve $3x+5=2x+13$.

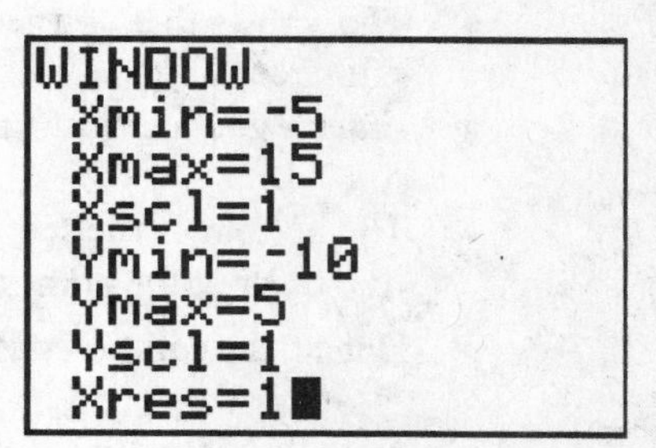

Find an approximate solution using TRACE

- First collect all terms to one side = 0.
- Subtract $2x+13$ from both sides to get the equation $3x+5-(2x+13)=0$
- Press [Y=] key directly below the screen.
- Use the [CLEAR] key to clear Y1 if needed.
- Enter Y1 $=3x+5-(2x+13)$.
- Press [GRAPH] to view the graph.
- Do you see the x-intercept of the line?
- If not, adjust the window until you do. Here is a suggested window and resulting graph.
- Press [TRACE] to show the equation on the graph.
- Use the right arrow to get close to the x-intercept. Note that the cursor is "close" to the x-intercept, but not exactly on it. We see that $x \approx 8$ when $y \approx 0$.

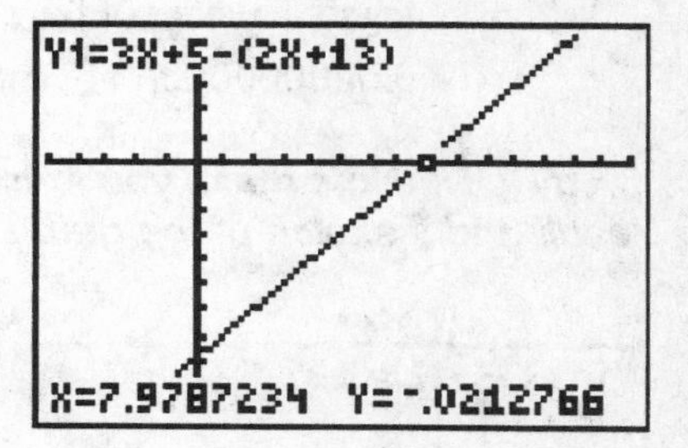

Find the exact solution using 2:zero in the CALC menu

- Press [2nd] [TRACE] (CALC) access the CALCULATE menu.
- Use the down arrow to scroll down to 2:zero and press [ENTER], or press the [2] key.
- The graph screen returns and asks for a Left bound.
- Position the cursor at a point close to the x-intercept, but to the left of it.
- You are asked this question because some curves to have more than one x-intercept.
- Linear functions always have just one x-intercept.
- You are defining the left side of the interval in which the x-intercept can be found.
- Press [ENTER].

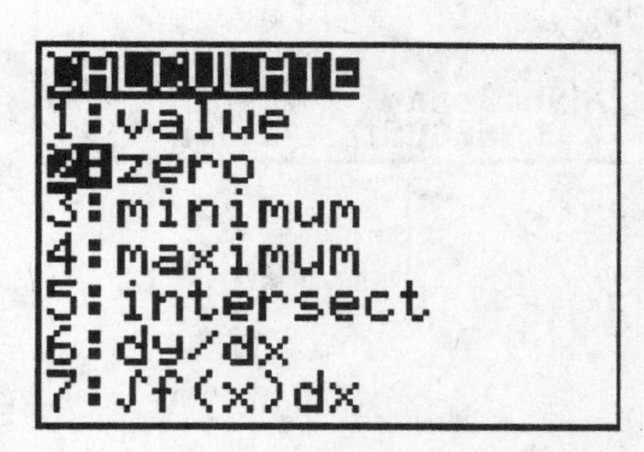

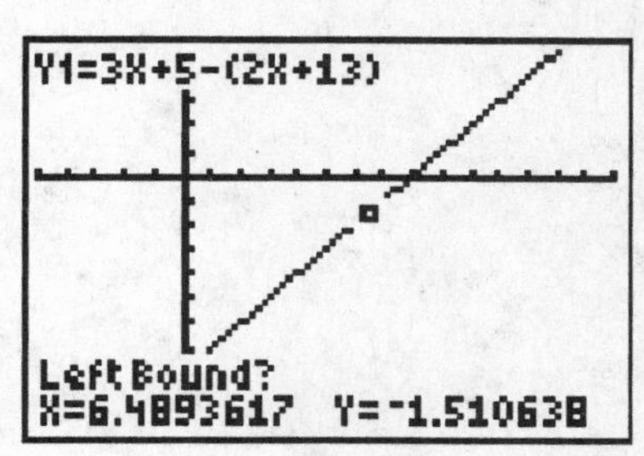

- Next, you see another screen that asks for the Right Bound?
- Position the cursor on a point close to the x-intercept, but to the right of it.
- You are defining the right side of the interval in which the x-intercept can be found.
- Press [ENTER].
- The next screen asks for a guess of the x value.
- You can enter a value or just press [ENTER].
- The arrows that you see above the line show the interval where the calculator will look for an x-intercept.
- Enter a guess or just Press [ENTER].

- The bottom of the screen shows that the x-intercept of this line occurs when $x = 8$ and $y = 0$, or at the point $(8, 0)$.
- Hence, the solution of the equation $3x + 5 = 2x + 13$ is $x = 8$.

- Note that sometimes you get a y-value of the zero displayed in scientific notation with a negative power of ten such as $5.34\ E - 10$. This number in standard notation is 0.000000000534, and is very close to zero.

***CAUTION:** Note that if you touch any other keys you lose the result. Be sure to write down the result and a sketch of the graph before you do anything else on the calculator.*

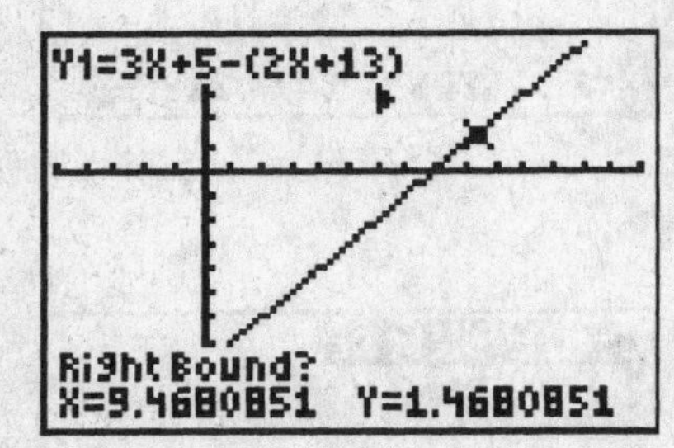

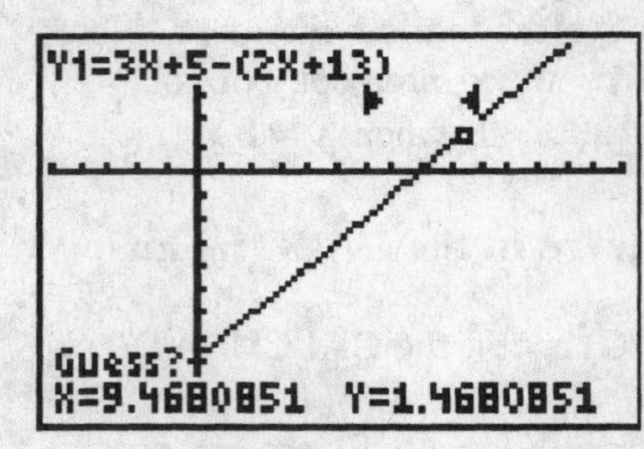

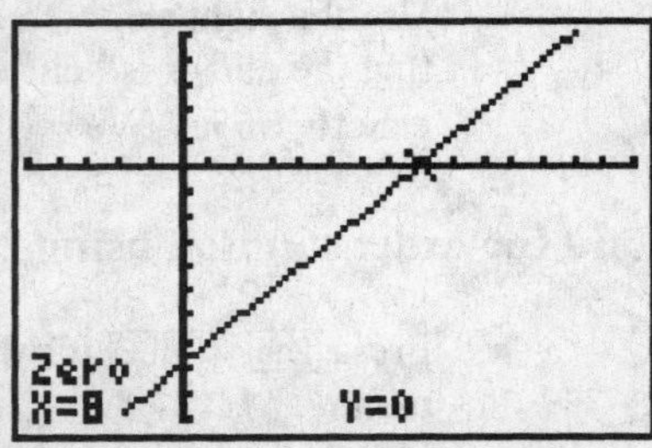

CHAPTER 8

LINEAR REGRESSION

LINEAR REGRESSION

Linear regression studies data that resembles a line and finds a linear equation that best fits or models the data. You must first enter the data as lists, format the plot to show a scatter plot of the data, find the linear regression equation, and finally graph both the scatter plot and the linear regression equation that models the data.

The problem: The consumption of cigarettes (in billions of cigarettes) is given in the table below.

Year	1980	1985	1990	1995	2000	2003
Year since 1980 (x)	0	5	10	15	20	23
Cigarettes in Billions (y)	632	615	525	487	461	449

- Create a scatter plot of the data points.
- Use linear regression to fit a line to the data.
- Graph the data and the regression line on the same screen.
- Predict the cigarette consumption in the year 2012 if the trend continues.

Create a Scatter Plot of Data Points

To create a scatter plot of data points, you must first select an appropriate window for the data. Then you must enter the data as lists, use the STAT menu to define Plot 1, Plot 2, or Plot 3, and finally graph the data points .

Set the Window

- Identify the lowest and highest values of x and y, then select a number about 5 or 10 more on each end.
- In this example, x goes from 0 to 23, and y goes from 449 to 632.
- The window on x as [-5, 30, 5] and on y as [400, 700, 50] includes these values.
- Press the WINDOW key.
- Set the window as shown on the right.

```
WINDOW
 Xmin=-5
 Xmax=30
 Xscl=5
 Ymin=400
 Ymax=700
 Yscl=50
 Xres=1
```

Clear Lists

EDIT CALC TESTS
1:Edit…
2:SortA(
3:SortD(
4:ClrList
5:SetUpEditor

- You may need to clear the lists before you enter the data.
- Press the [STAT] key.
- Use the down arrow to select 4: ClrList. This pastes ClrList on the Home Screen.
- Next, you must tell the calculator which list to clear.
- Press [2nd] [1] (L_1). This puts L_1 next to ClrList.
- Press [ENTER] to clear L_1.
- You should see the word Done.
- Repeat the line by pressing [2nd] [ENTER] (ENTRY).
- Move the cursor over L_1.
- Press [2nd] [2] (L_2) to replace L_1 with L_2 .
- Press [ENTER] to clear L_2.

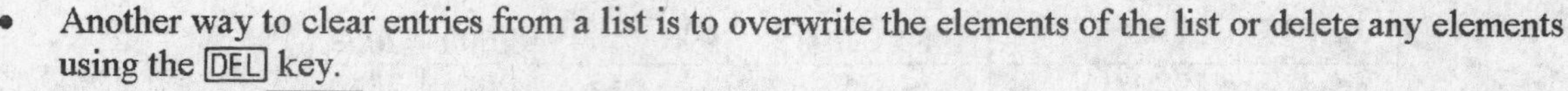

- Another way to clear entries from a list is to overwrite the elements of the list or delete any elements using the [DEL] key.
- Note that the [CLEAR] key does not erase a highlighted entry.

ClrList L1
Done

- An entire list can be cleared in the STAT menu by pressing the up arrow until the top of the list is highlighted.
- Press [CLEAR] [ENTER] in this position to clear the contents of the entire list.

- If you press the [DEL] key when the column heading is highlighted, the list is removed from the STAT editor, but the list will still possess its contents.
- The list name can be returned to the editor.
- Press [2nd] [DEL] (INS) and type the list name.

Enter the Data Points

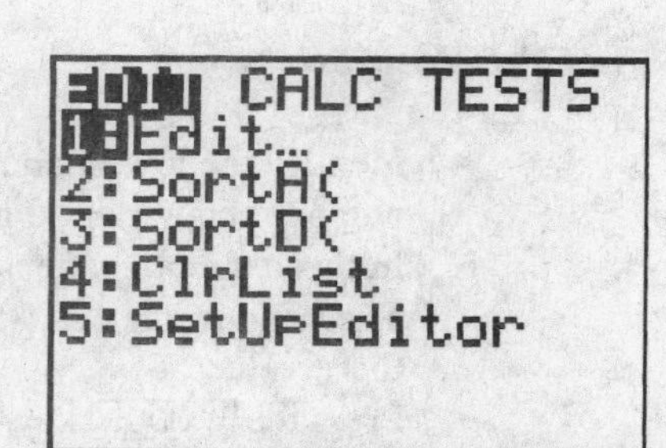

- Press the [STAT] key.
- Press [1] to select 1:Edit or move the cursor to 1:Edit and press [ENTER]. This brings you to the screen where you can enter the data in the lists.
- You may need to clear the lists. See directions above about clearing lists.
- Enter the Years from 1980-values in list L_1.
- Enter the Cigarettes in Billions-values in L_2.
- Use the down arrow to move from line to line.
- Press [2nd] [MODE] (QUIT) to return to the Home screen.

L1	L2	L3 1
0	632	------
5	615	
10	525	
15	487	
20	461	
23	449	
------	------	

L1 ={0,5,10,15,2…

Turn on a Plot

- Go back to the Y= menu and turn off or clear all the Y= functions, namely the ones that have a dark rectangle over the equal sign.
- To turn off a function, move the cursor over the *equal* sign of the function you wish to turn off.
- Press [ENTER] to remove the highlighted (dark) rectangle over the equal sign.
- Turn on Plot 1.
- On the graph screen, use the up arrow to go to Plot 1.
- Press [ENTER] to highlight it. Plot 1 is now *on*.
- Press [GRAPH] to see the points.
- Note that Plot 1 must be defined to graph from lists L_1 and L_2. If not, format Plot 1 using STAT PLOT as described below.
- If you see something else, like a histogram, you must format Plot 1 using STAT PLOT as described below.
- Adjust the window if you wish to see the points more clearly.

Use STAT PLOT to format the graph of data points

- Press [2nd] [Y=] to access STAT PLOT.
- Press [ENTER] to select 1: Plot 1.
- Position the cursor over the word ON.
- Press [ENTER].
- Move the cursor over the first type of plot. Select the first setting, scatter plot.
- Press [ENTER].
- On the same screen select
  XList: L_1 or whichever list has the x-data
  YList: L_2 or whichever list has the y-data
  Mark: any one you wish
- Press the [GRAPH] key to see the scatter plot.
- Use [TRACE] and the right arrow to move from point to point.
- The x and y coordinates of each point are displayed at the bottom of the screen.
- P1:L1,L2 is displayed in the upper left hand corner of the screen. This tells you that you are on Plot 1, the x's are from list L_1, and the y's are from list L_2.

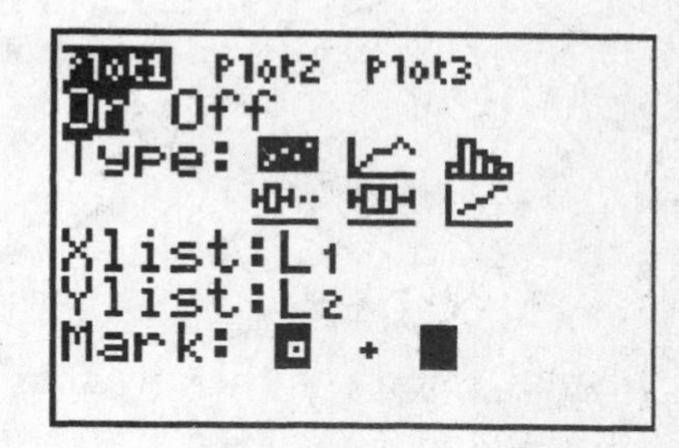

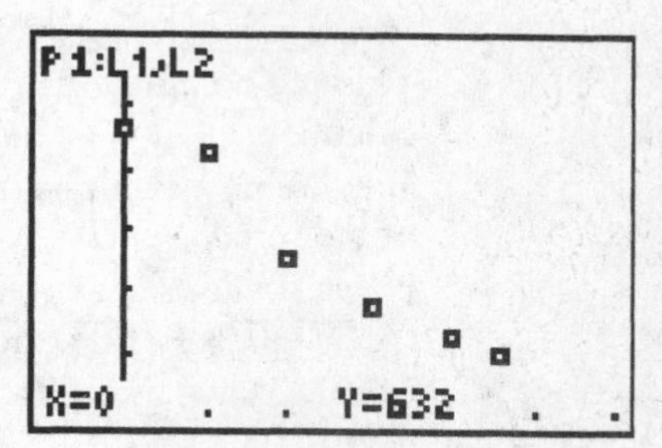

Find and Graph the Linear Regression Equation that is the "Best Fit" for the Data.

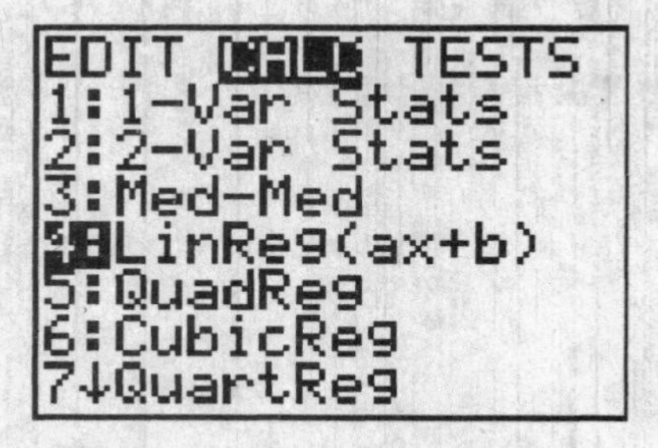

- The data is already entered into L1 and L2, and Plot 1 is formatted for scatter plot.
- Turn off or clear Y1, Y2, and Y3.
- Press the [STAT] key and use the right arrow to go to CALC.
- Press [4] to select 4:LinReg(ax+b). This puts LinReg(ax+b) on the Home screen.
- Press [ENTER] to see the values of a and b as shown.
- If you do not see r^2 and r, see the directions below on how to turn Diagnostics On.
- Hence, the equation of the line through the data points is $y = -8.646055437x + 633.3603412$.
- Use the [MODE] key and Float 3 to round these values to 3 decimals as shown to see the equation more easily.
- Press [CLEAR] to return to the Home screen. Press [ENTER] to see the Linear Regression equation in 3 decimal places, $y = -8.646x + 633.360$.

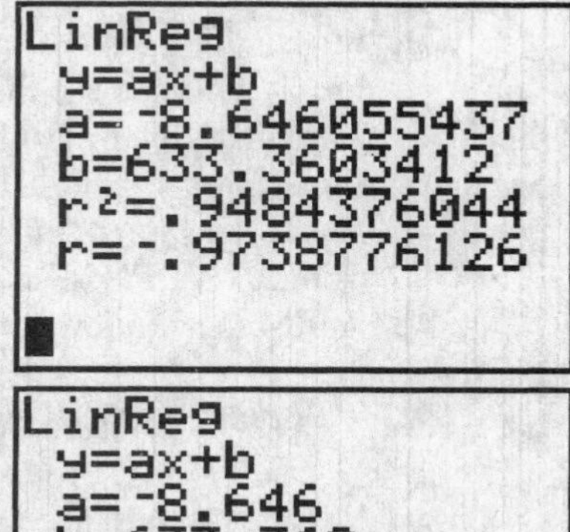

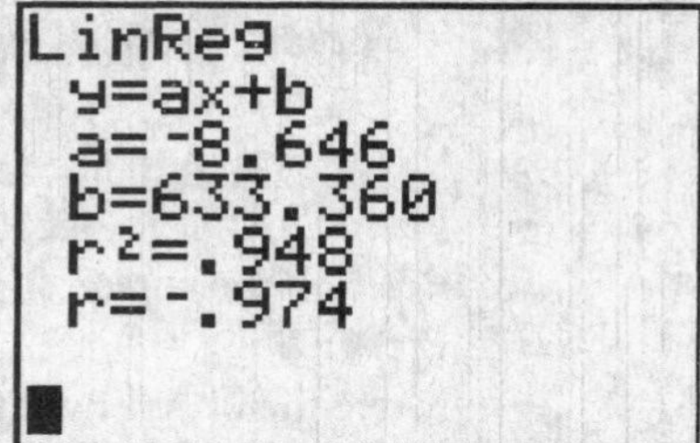

Display r and r^2 by setting DiagnosticOn

You may not see the values for the linear correlation coefficient *r*. To do so, you must go to the Catalog and set DiagnosticOn.

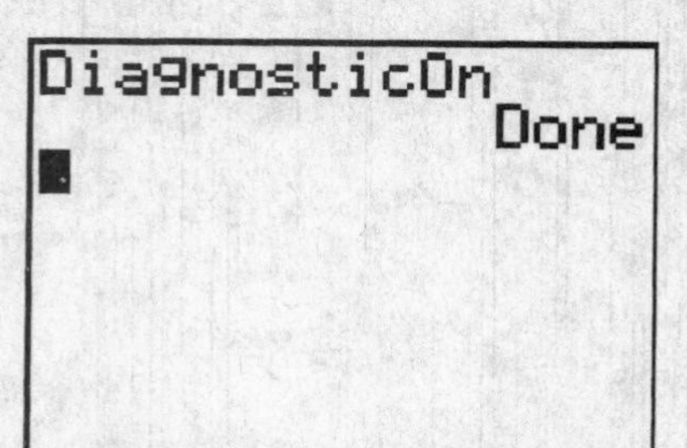

- Press [2nd] [0] to access the catalog.
- This is a complete listing of all the functions available on the calculator.
- You are now in alpha mode (see the A in the upper right hand corner of the screen.
- If you do not see an A in the upper right corner of the screen, press the [ALPHA] key first, then press the [x^{-1}] key to jump to the D's.
- Use the down arrow to find DiagnosticOn. Be sure that the arrow is next to it.
- Press [ENTER] twice until you see the word Done on the Home Screen.
- Redo finding the linear equation. The screen will now also include *r* and r^2.

DiagnosticOn
Done

Graph the Linear Regression Equation and Scatter Plot from the Home Screen

- Set the mode to the word FLOAT to get more accurate values.
- Turn off all Y= functions. Be sure that Plot 1 is on.
- Press [2nd] [MODE] (QUIT) to go to the Home screen.
- Press the [STAT] key and use the right arrow to go to CALC.
- Press [4] to select 4:LinReg(ax+b). This puts LinReg(ax+b) on the Home screen.
- Enter the names of the lists to use for the linear regression equation.
- Press [2nd] [1] (L1) comma [,] [2nd] [2] (L2) comma [,] to enter L1 and L2.

```
EDIT CALC TESTS
1:1-Var Stats
2:2-Var Stats
3:Med-Med
4:LinReg(ax+b)
5:QuadReg
6:CubicReg
7↓QuartReg
```

- Next, you must name the Y= function where you wish to put the linear regression equation.
- If you wish to put the linear regression equation in the Y= menu as Y1, you must find Y1 in the VARS menu.
- Press the [VARS] key and use the right arrow to go to Y-VARS.
- Select 1:Function.
- On the Function screen, select the position in the Y=menu you wish to use for the regression equation, say Y1.

```
VARS Y-VARS
1:Function...
2:Parametric...
3:Polar...
4:On/Off...
```

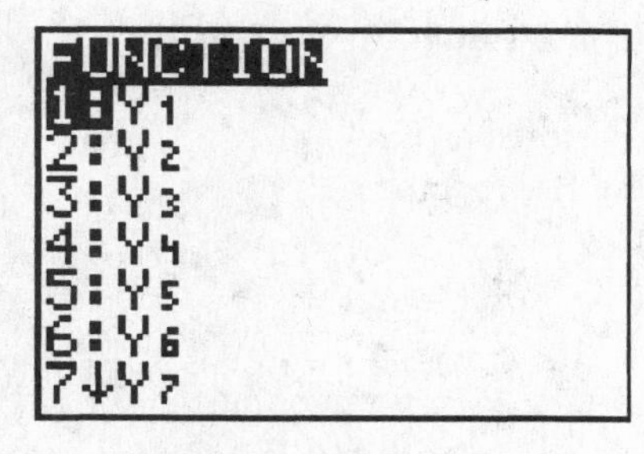

- Press [ENTER] to see the linear regression information on the Home screen.
- Press [Y=] to see it in the Y= menu.
- Press [GRAPH] to see the scatter plot and the linear regression equation on the same screen.

```
LinReg
 y=ax+b
 a=-8.646055437
 b=633.3603412
 r²=.9484376044
 r=-.9738776126
```

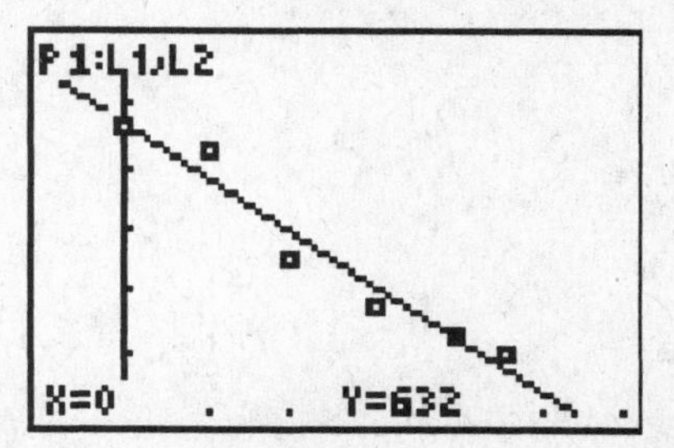

Graph the Linear Regression Equation and Scatter Plot from the Graph Screen

- Turn off all Y= functions.
- Be sure that Plot 1 is on.
- Move the cursor to the Y= line you wish to use, say Y1.
- Press the [VARS] key.
- Select 5:Statistics.
- Move the right arrow twice to see the EQ menu.
- Select 1:RegEQ.
- Press [ENTER]. This puts the regression equation into Y1 on the Y= screen.
- Press the [GRAPH] key to see the regression equation and the scatter plot on the same screen.
- Press [TRACE] to trace the graphs.
- Use the up or down arrow to toggle between the scatter plot and the linear regression equation.

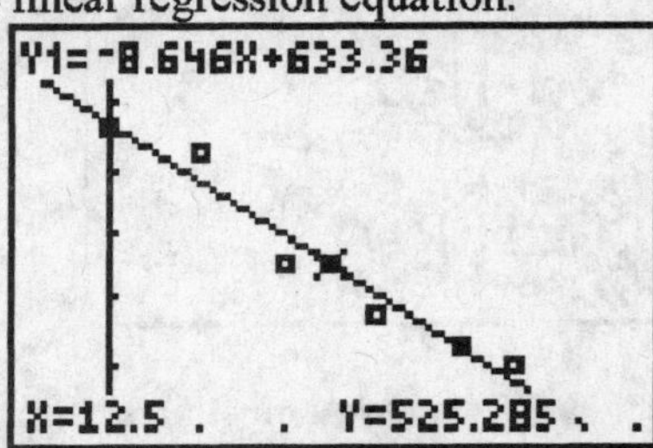

Use the Linear Regression Equation to predict future cigarette usage

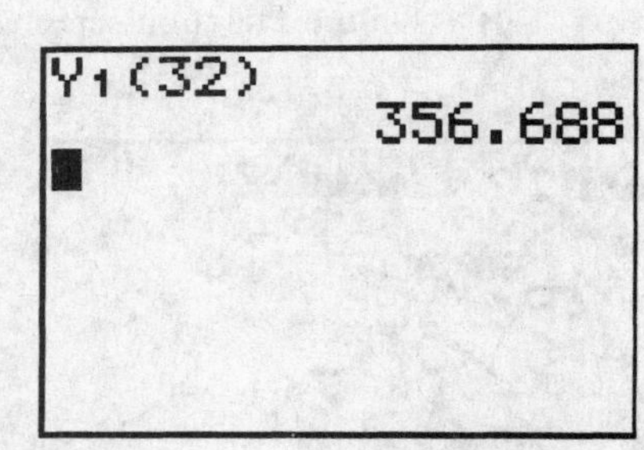

- The year 2012 is 32 years after 1980.
- You must find the value of the linear regression equation when $x = 32$.
- Recall that the regression equation is stored as Y1.
- Use the [VARS] menu to put Y1 on the Home screen.
- Press the [VARS] key.
- Use the right arrow to go to Y-VARS.
- Select 1:Function.
- Select 1:Y1.
- Press [(][3][2][)] to see Y1(32).
- Press [ENTER] to see the predicted value, 356.688, which means that approximately 357 billion people will be smoking cigarettes in the year 2012.

CHAPTER 9

SYSTEMS OF LINEAR EQUATIONS

SOLVE A SYSTEM OF LINEAR EQUATIONS GRAPHICALLY

Solve the system $\begin{matrix} y=-x-1 \\ 4x-3y=24 \end{matrix}$ by graphing.

Find the Exact Coordinates of the Intersection Point

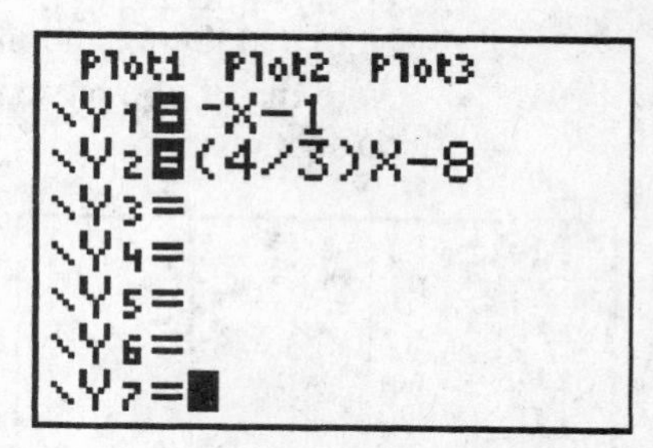

- Turn off or clear any equations or plots.
- Solve each equation for y and enter them as Y1 and Y2.
- Enter Y1=$-x-1$ and Y2=$(4/3)x-8$.
- Use the Standard window.
- Press [ZOOM], select 6:ZStandard.
- Press the [GRAPH] key and then [TRACE] to see the graph.
- Use the arrow to move the cursor close to the intersection point.
- Press [2nd] [TRACE] (CALC).
- Press the [5] key, or cursor down to 5:intersect and press [ENTER].
- You must select the *First curve*.
- Note that the equation the cursor is on is shown at the top left of the screen.
- Press [ENTER].
- Next you must select a *Second curve*.
- Use the up or down arrow key to move to the other line.
- Press [ENTER].
- Next, you must enter a *Guess* or just press [ENTER] to see the coordinates of the Intersection at the bottom of the screen. The intersection of this system of linear equations is the point $(3,-4)$.

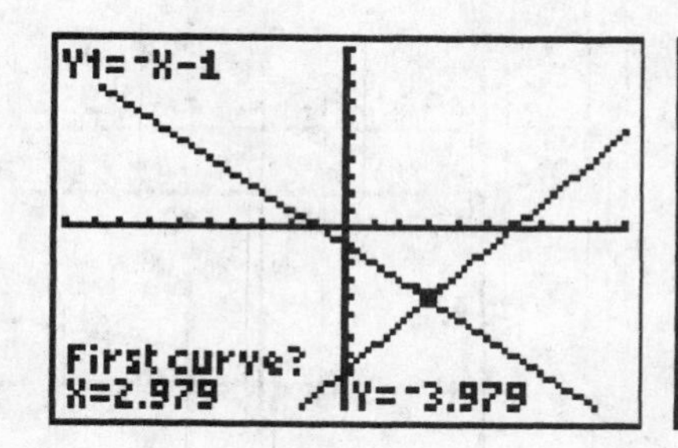

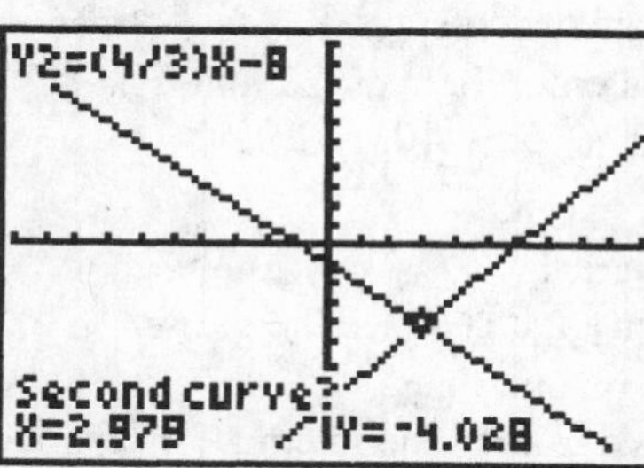

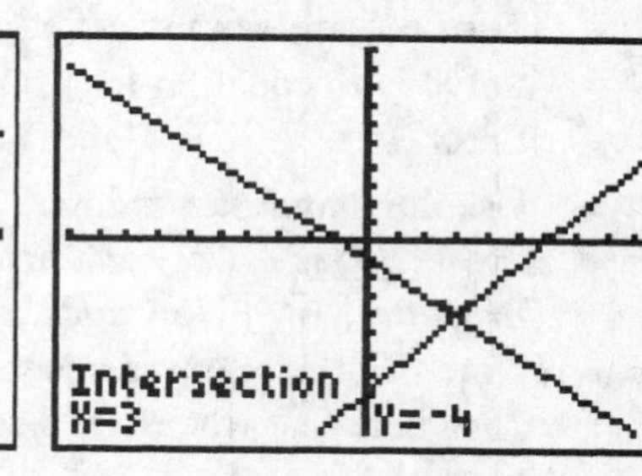

Solve a System of Parallel Lines with no Solution

Solve the system $\begin{matrix} 5x-2y=4 \\ -10x+4y=7 \end{matrix}$ by graphing.

- Turn off or clear any equations or plots.
- Solve each equation for y and enter them as Y1 and Y2.
- Enter $\text{Y1}=(5/2)x-2$ and $\text{Y2}=(10/4)x+7/4$.
- Use the Standard window.
- Press ZOOM, select 6:ZStandard.
- Press the GRAPH key and then TRACE to see the graph.
- Note that there is no intersection point visible. Try a few more window settings.
- Note that if you try to use CALC 5:intersect to find the intersection, you get an error message that there is no sign change.

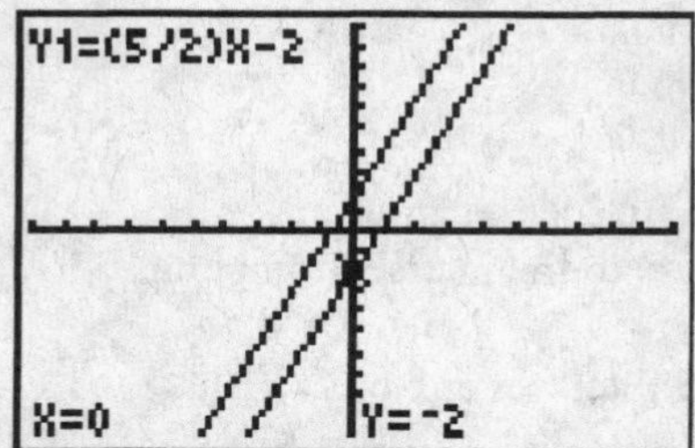

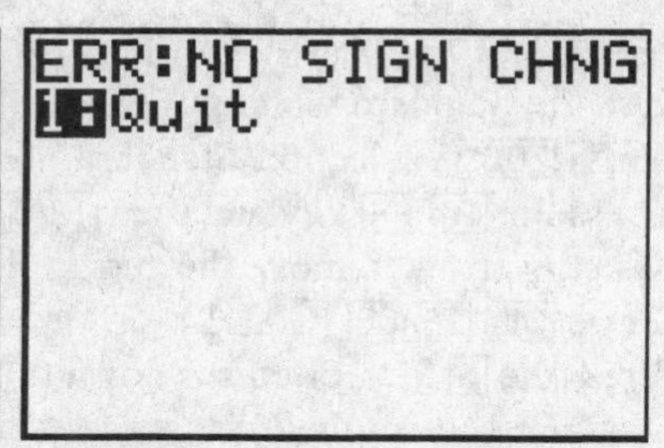

Solve a System with an Infinite Number of Solutions

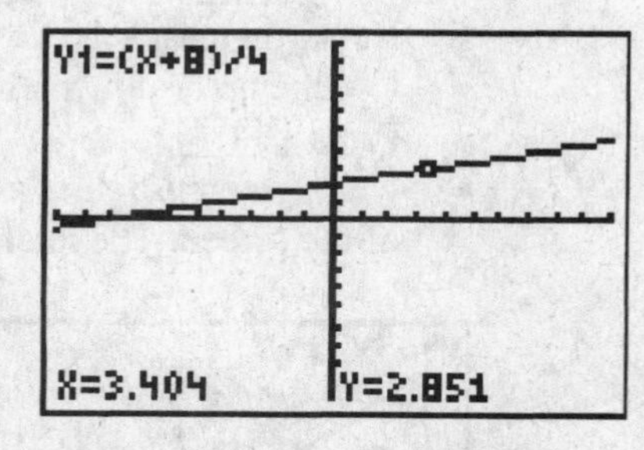

Solve the system $\begin{matrix} x=4y-8 \\ 5x-20y=-40 \end{matrix}$ by graphing.

- Turn off or clear any equations or plots.
- Solve each equation for y and enter them as Y1 and Y2.
- Enter $\text{Y1}=(x+8)/4$ and $\text{Y2}=(-5x-40)/-20$.
- Use the Standard window.
- Press ZOOM, select 6:ZStandard.
- Press the GRAPH key and then TRACE to see the graph.
- Note that the graph seems to be one line.
- Also note that when you move to the second equation, the coordinates at the bottom of the screen do not change.

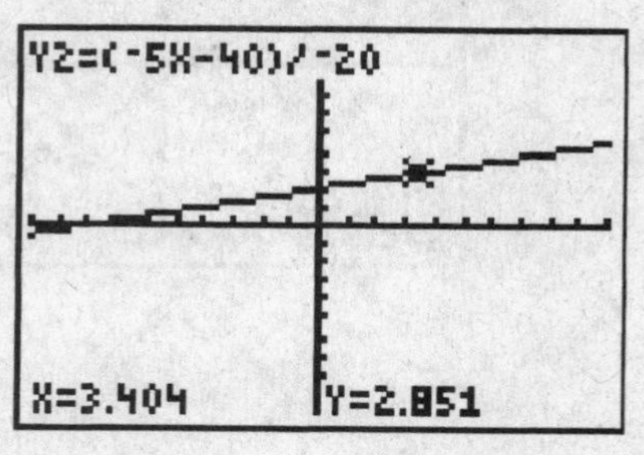

- Let's see what happens when we ask for the intersection.
- Press [2nd] [TRACE] (CALC).
- Press the [5] key, or cursor down to 5:intersect and press [ENTER].

- You must select the *First curve*.
- Note that the equation the cursor is on is shown at the top left of the screen.
- Press [ENTER].

- Next you must select a *Second curve.*
- Use the up or down arrow key to move to the other line.
- Press [ENTER].

- Next, you must enter a *Guess* or just press [ENTER].

- Note that as you move the cursor anywhere on the curve and ask for an intersection point, you get another intersection point.
- Try it at least three times.

- This result means that every point of the first curve is also a solution to the second curve, hence there are an infinite number of intersection points.

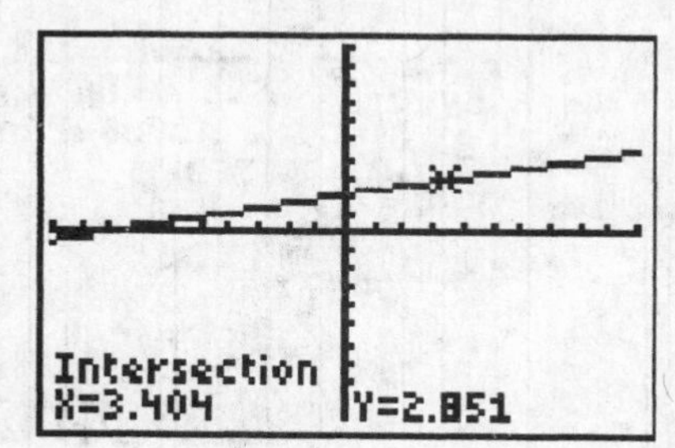

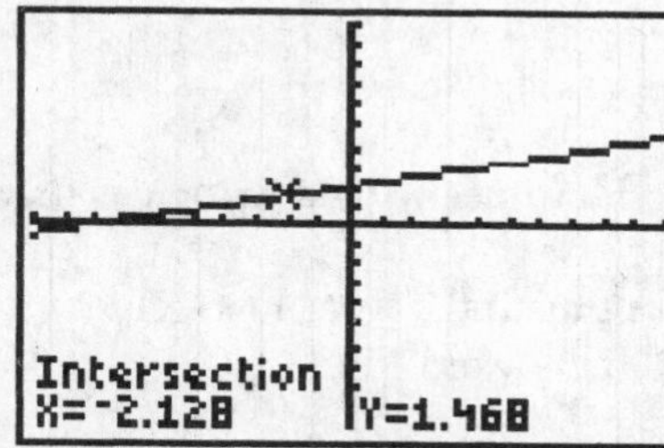

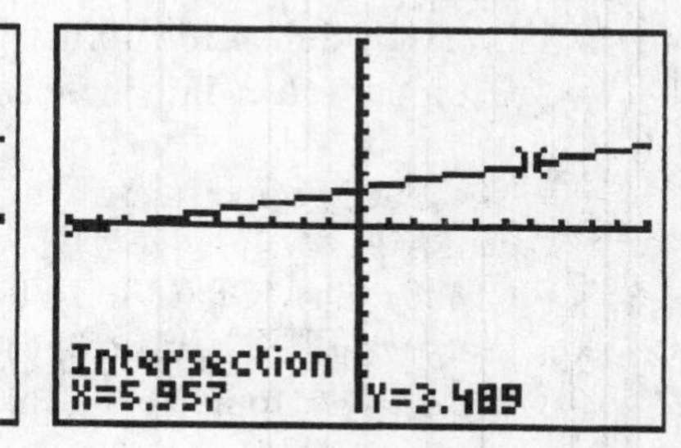

SOLVE SYSTEMS OF LINEAR EQUATIONS BY MATRICES

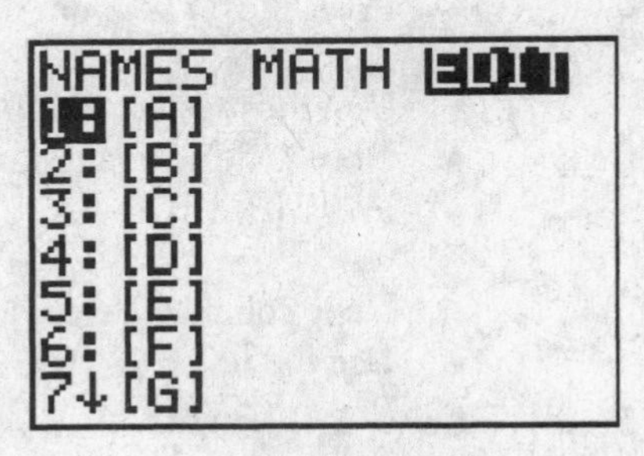

Solve the system of equations $\begin{cases} 3x+y+2z=31 \\ x+y+2z=19 \\ x+3y+2z=25 \end{cases}$ with matrices.

First represent the system as the augmented matrix $\begin{bmatrix} 3 & 1 & 2 & 31 \\ 1 & 1 & 2 & 19 \\ 1 & 3 & 2 & 25 \end{bmatrix}$.

Enter the Matrix

- Press [2nd] [x^{-1}] to access the MATRIX menu.
- Use the right arrow to go to EDIT.
- Move the cursor to 1: [A] and press [ENTER].
- Note that if you used this matrix name before, it will have a dimension next to it.
- Change the dimension of matrix on the next screen.
- Enter the dimension of matrix A as 3 x 4.
- Enter the values into the matrix as shown.

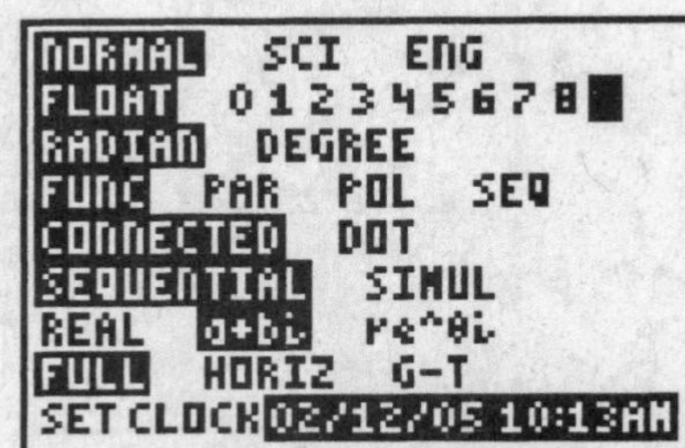

- The matrix screen shows that the values will be written as three place decimals.
- Change the MODE FLOAT setting to the word FLOAT.
- Press the [MODE] key, go down to the word FLOAT.
- Press [ENTER] to make the change.

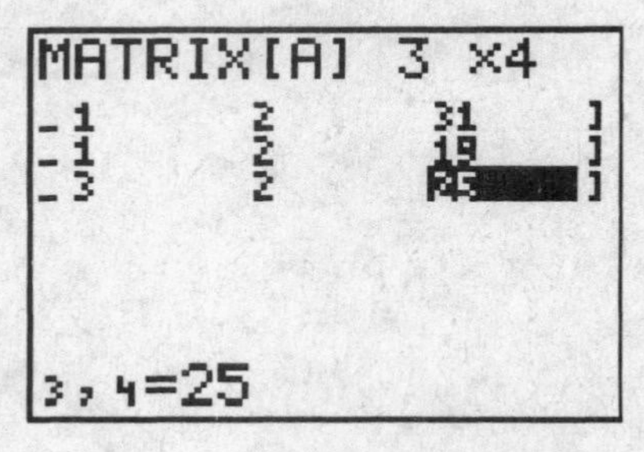

- Go back to the MATRX EDIT screen to enter the matrix.
- Note that the position is given at the bottom of the screen as 1,1=0 etc.
- The first number is the row, the second number is the column, the third number is the current value of the position.
- Press [ENTER] for each number entered in the matrix.

- This matrix needs two screens.
- Use the right arrow to see the last column.
- You can enter fractions in fraction form, but the calculator shows the decimal equivalent.
- Press [2nd] [MODE] to QUIT and return to the Home screen.

View the Matrix on the Home Screen using Matrix Names

- Press [2nd] [x^{-1}] to access the MATRIX menu.
- You are in the NAMES menu.
- Move the cursor to 1: [A] and press [ENTER].
- This will put [A] on the Home screen.
- Press [ENTER] again to view the matrix on the Home screen.

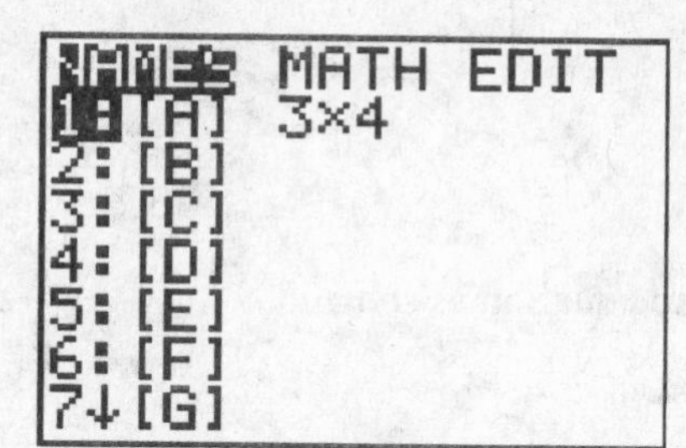

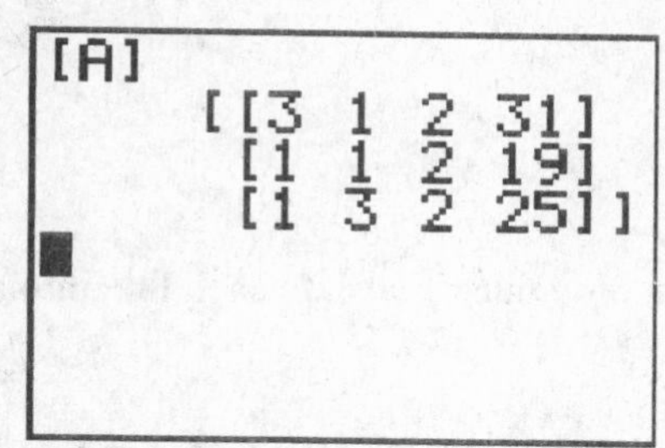

Find the row-echelon form of the matrix, ref ([A])

- Press [2nd] [x^{-1}] to access the MATRIX menu.
- Use the right arrow to go to MATH.
- Use the down arrow to select A: ref(.
- Press [ENTER].
- This puts ref(on the Home screen.
- To ensure that the results will be shown as fractions, press the [MATH] key, select 1:➔FRAC.
- Enter the name of the matrix you wish to use in the parentheses.
- You must use MATRIX NAMES to enter the name of the matrix. Otherwise you get an ERR: SYNTAX message.
- Press [ENTER].
- Use the right arrow to see the entire matrix.

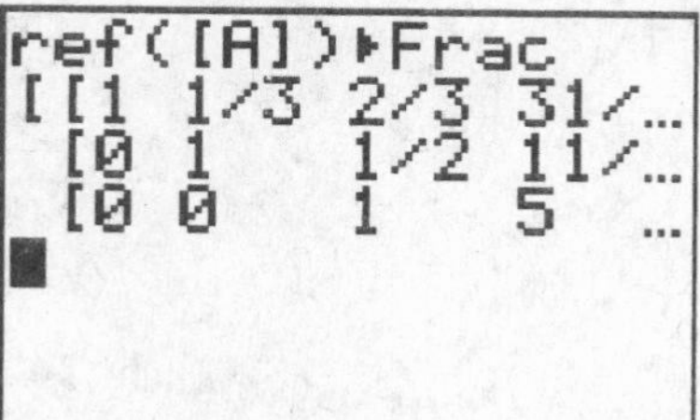

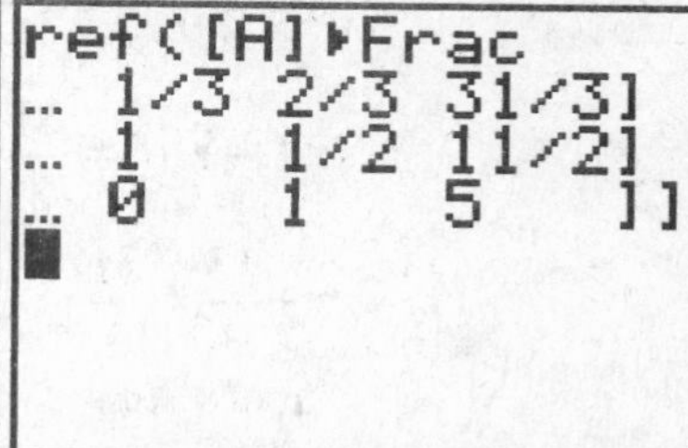

Find the solution from ref form of the matrix

- Record this matrix on your own paper as $\begin{bmatrix} 1 & \frac{1}{3} & \frac{2}{3} & \frac{31}{3} \\ 0 & 1 & \frac{1}{2} & \frac{11}{2} \\ 0 & 0 & 1 & 5 \end{bmatrix}$.

- To find the solution, you must translate each line into an equation as shown.

$$x+\frac{1}{3}y+\frac{2}{3}z=\frac{31}{3}$$
$$y+\frac{1}{2}z=\frac{11}{2}$$
$$z=5$$

- Next, start with the last line, which says that $z=5$.
- Substitute $z=5$ into the second equation $y+\frac{1}{2}z=\frac{11}{2}$ and solve for y.
- Solve for y

$$y+\frac{1}{2}(5)=\frac{11}{2}$$
$$y=\frac{11}{2}-\frac{5}{2}=\frac{6}{2}$$
$$y=3$$

- Finally, substitute $z=5$ and $y=3$ into the first equation and solve for x.
- Solve for x

$$x+\frac{1}{3}y+\frac{2}{3}z=\frac{31}{3}$$

$$x+\frac{1}{3}(3)+\frac{2}{3}(5)=\frac{31}{3}$$
$$x+\frac{3}{3}+\frac{10}{3}=\frac{31}{3}$$
$$x+\frac{13}{3}=\frac{31}{3}$$
$$x=\frac{31}{3}-\frac{13}{3}=\frac{18}{3}$$
$$x=6$$

- Hence, the solution of the system $\begin{aligned} 3x+y+2z&=31 \\ x+y+2z&=19 \\ x+3y+2z&=25 \end{aligned}$ is $x=6,\ y=3,\ z=5$

Find the reduced row-echelon form of the matrix, rref ([A])

- Press [2nd] [x^{-1}] to access the MATRIX menu.
- Use the right arrow to go to MATH.
- Use the down arrow to select B: rref(.
- Press [ENTER].
- This puts rref(on the Home screen.

- Enter the name of the matrix you wish to use in the parentheses.
- You must use MATRIX NAMES to enter the name of the matrix. Otherwise you get an ERR: DATA TYPE message.
- Press [ENTER] to see the final matrix.

- This says that the solution to this system of linear equations is $x = 6, y = 3, z = 5$, or the point $(6,3,5)$

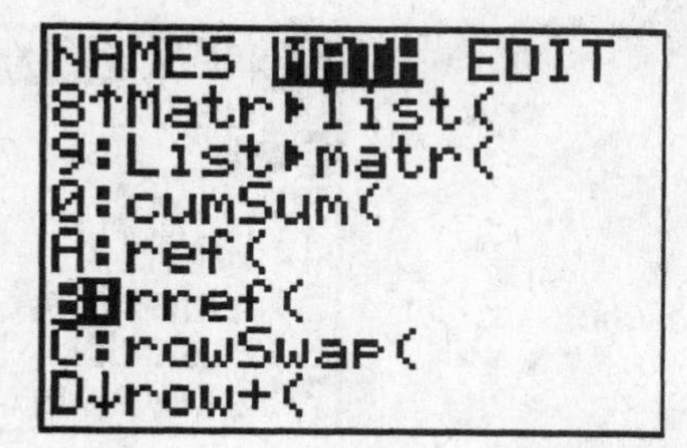

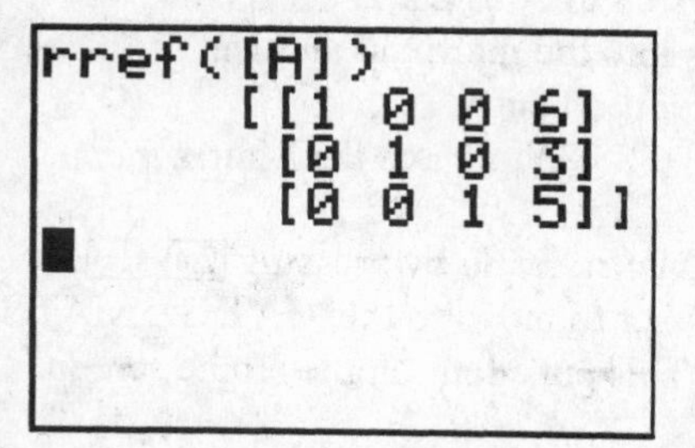

SOLVE A SYSTEM OF LINEAR EQUATIONS BY DETERMINANTS

Find the Determinant of a Matrix

To find the determinant of a matrix, first enter the matrix, then use the det function in the MATRIX MATH menu.

Find the determinant of the matrix $\begin{bmatrix} 5 & 6 \\ 7 & 3 \end{bmatrix}$.

- Press [2nd] [x^{-1}] to access the MATRIX menu.
- Use the right arrow to go to EDIT.
- Move the cursor to 1: [A] and press [ENTER].
- Change the dimension of matrix on the next screen.

- Enter the dimension of A as 2 x 2.
- Enter the values into the matrix as shown.
- Press [ENTER] after each entry.
- Press [2nd] [MODE] (QUIT) to leave the Matrix menu.

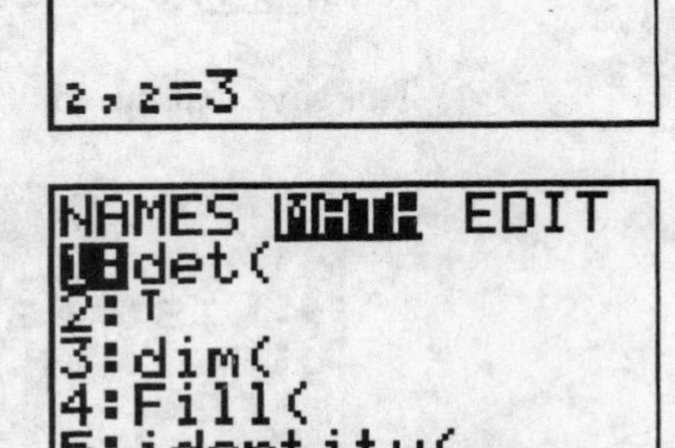

- Go back to the Matrix menu by pressing [2nd] [x^{-1}].
- Use the right cursor to move to the MATRIX MATH menu.
- Select 1:det(. This puts det(on the Home screen.

- Use MATRIX NAMES in the Matrix menu to enter the name of the matrix to be used in the determinant function.
- Press [ENTER] twice to see the value of the determinant of Matrix A as -27.

Use Cramer's Rule to Solve a System of Linear Equations in Two Variables

If $\begin{array}{l} a_1x+b_1y=c_1 \\ a_2x+b_2y=c_2 \end{array}$ then $x=\dfrac{\begin{vmatrix} c_1 & b_1 \\ c_2 & b_2 \end{vmatrix}}{\begin{vmatrix} a_1 & b_1 \\ a_2 & b_2 \end{vmatrix}}=\dfrac{D_x}{D}$ and $y=\dfrac{\begin{vmatrix} a_1 & c_1 \\ a_2 & c_2 \end{vmatrix}}{\begin{vmatrix} a_1 & b_1 \\ a_2 & b_2 \end{vmatrix}}=\dfrac{D_y}{D}$ where $\begin{vmatrix} a_1 & b_1 \\ a_2 & b_2 \end{vmatrix}\neq 0$

Use Cramer's Rule to solve the system: $\begin{array}{l} 5x-4y=2 \\ 6x-5y=1 \end{array}$.

Enter the three matrices D, D_x, D_y **as matrices** $[A]$, $[B]$, $[C]$

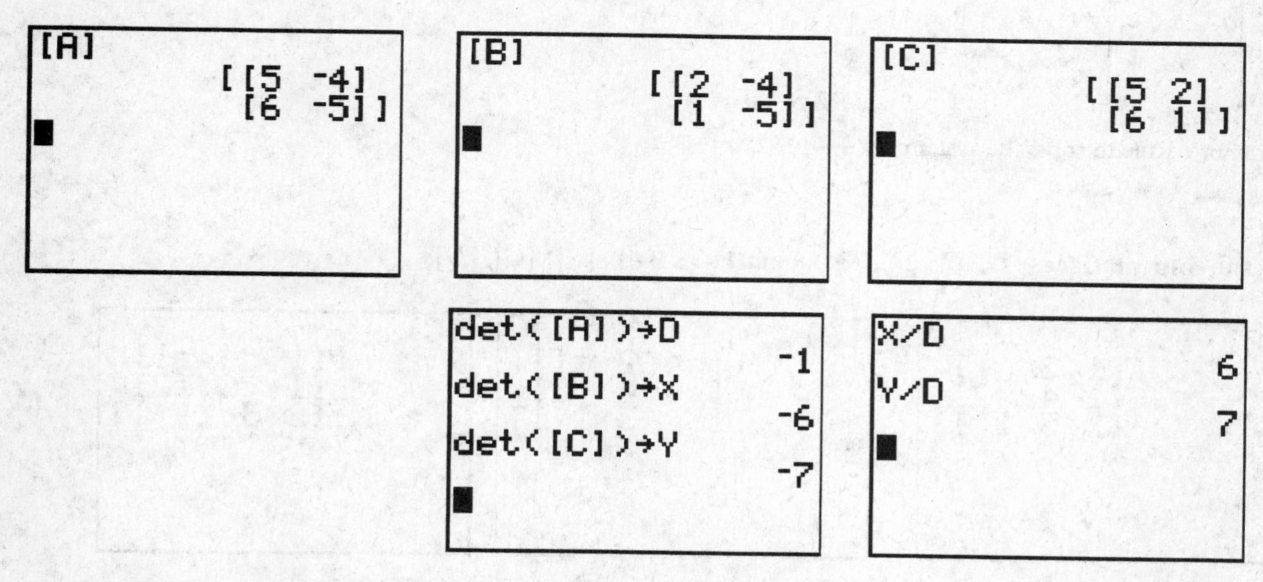

- Follow the directions for entering a matrix using MATRIX EDIT.
- Enter the denominator D as matrix $[A]$, D_x as matrix $[B]$, and D_y as matrix $[C]$.
- Press 2nd x^{-1} (MATRIX) and go back to the MATRIX EDIT menu after entering each matrix.
- Press 2nd MODE (QUIT) to return to the Home screen.
- You can evaluate each determinant separately and then do the division as shown above.
- Store the results as shown, then do the division of the stored results.
- State that the solution to this system is the point $(6, 7)$.
- Or you can do the division of the determinants on one line as shown on the right.
- You must use MATRIX MATH and MATRIX NAMES for this computation.
- State that the solution to this system is the point $(6, 7)$.

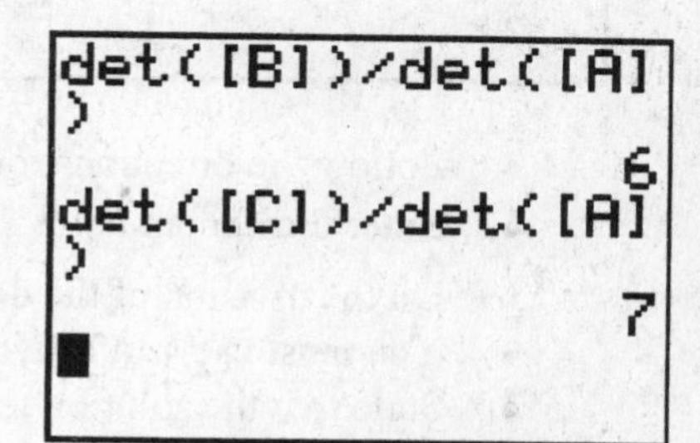

Use Cramer's Rule to solve a System of Linear Equations in Three Variables

If $\begin{aligned} a_1x+b_1y+c_1z&=d_1 \\ a_2x+b_2y+c_2z&=d_2 \\ a_3x+b_3y+c_3z&=d_3 \end{aligned}$ then $x=\dfrac{\begin{vmatrix} d_1 & b_1 & c_1 \\ d_2 & b_2 & c_2 \\ d_3 & b_3 & c_3 \end{vmatrix}}{\begin{vmatrix} a_1 & b_1 & c_1 \\ a_2 & b_2 & c_2 \\ a_3 & b_3 & c_3 \end{vmatrix}}=\dfrac{D_x}{D}$, $y=\dfrac{\begin{vmatrix} a_1 & d_1 & c_1 \\ a_2 & d_2 & c_2 \\ a_3 & d_3 & c_3 \end{vmatrix}}{\begin{vmatrix} a_1 & b_1 & c_1 \\ a_2 & b_2 & c_2 \\ a_3 & b_3 & c_3 \end{vmatrix}}=\dfrac{D_y}{D}$ $z=\dfrac{\begin{vmatrix} a_1 & b_1 & d_1 \\ a_2 & b_2 & d_2 \\ a_3 & b_3 & d_3 \end{vmatrix}}{\begin{vmatrix} a_1 & b_1 & c_1 \\ a_2 & b_2 & c_2 \\ a_3 & b_3 & c_3 \end{vmatrix}}=\dfrac{D_z}{D}$

where $\begin{vmatrix} a_1 & b_1 & c_1 \\ a_2 & b_2 & c_2 \\ a_3 & b_3 & c_3 \end{vmatrix} \neq 0$.

Use Cramer's Rule to solve the system: $\begin{aligned} x+2y-\ z&=-4 \\ x+4y-2z&=-6 \\ 2x+3y+\ z&=\ 3 \end{aligned}$.

Enter the four matrices $D,\ D_x,\ D_y,\ D_z$ **as matrices** $[A],\ [B],\ [C],\ [D]$

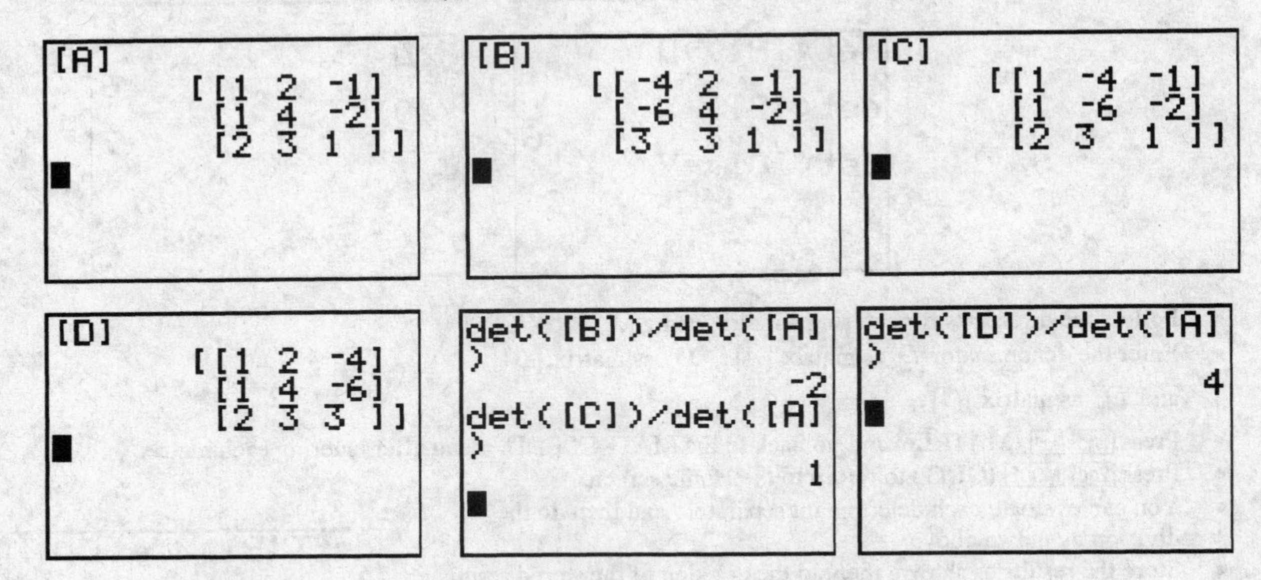

- Follow the directions for entering a matrix using MATRIX EDIT.
- Enter the denominator D as matrix $[A]$, D_x as matrix $[B]$, D_y as matrix $[C]$, and D_z as matrix $[D]$.
- Do the division of the determinants on one line as shown.
- You must use MATRIX MATH and MATRIX NAMES for this computation.
- State that the solution to this system is the ordered triple $(-2, 1, 4)$.

CHAPTER 10

POLYNOMIALS

SCIENTIFIC NOTATION

You may compute using scientific notation in NORMAL MODE or in SCI MODE. If you wish, you may use the 10^x function found above the [LOG] key, or you may use the EE function found above the comma [,] key. In NORMAL mode, all results will be shown in standard notation. In SCI mode, all results will be shown in scientific notation. To convert a scientific notation back to normal, you must change the mode back to NORMAL.

Compute using EE [2nd] [,]

Compute 2.5×10^3. Express the result in Normal notation.

- Press the [MODE] key.
- Highlight the word NORMAL on the first line.
- Press [ENTER] to set the Normal mode.
- Press [CLEAR] to return to the Home screen.
- Press [2] [.] [5] [2nd] [,] [3] to put 2.5 E 3 on the Home screen.
- Press [ENTER] to see the normal notation for this number, 2500.

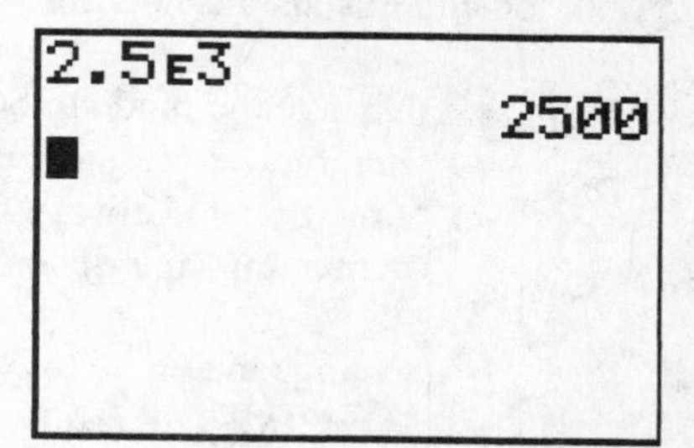

Compute 2.5×10^3. Express the result in Scientific notation.

- Press the [MODE] key. Highlight the word SCI on the first line.
- Press [ENTER] to set the SCI mode.
- Press [CLEAR] to return to the Home screen.
- Press [2] [.] [5] [2nd] [,] [3] to put 2.5 E 3 on the Home screen.
- Press [ENTER] to see the scientific notation for this number, 2.5 E 3.

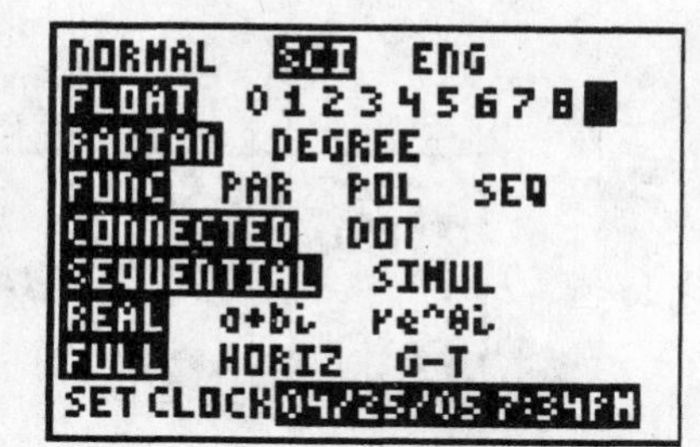

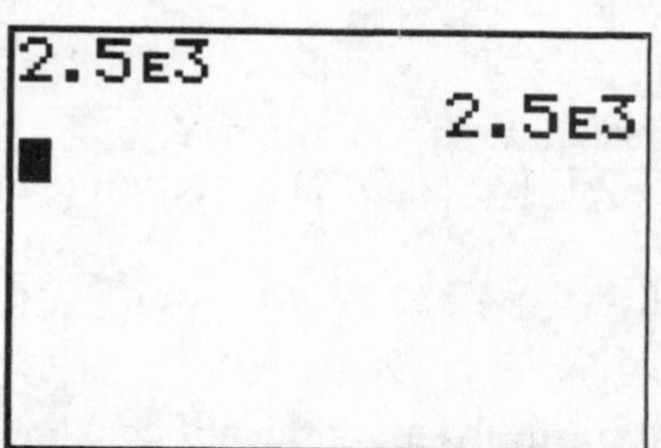

Compute $(9.6\times10^{6})(4\times10^{-8})$ using EE. Show the results in both scientific mode and in normal mode.

- Press the MODE key. Select SCI for scientific notation. Press ENTER.
- Press CLEAR or 2nd MODE (QUIT) to return to the Home screen.
- Enter (9.6 E 6)(4 E -8). Use 2nd comma , to enter E.
- Press ENTER to see the result in scientific notation.

- Change the mode back to normal to see the result in standard notation.
- Press ENTER to see the result in normal notation. It is not necessary to re-enter expression.

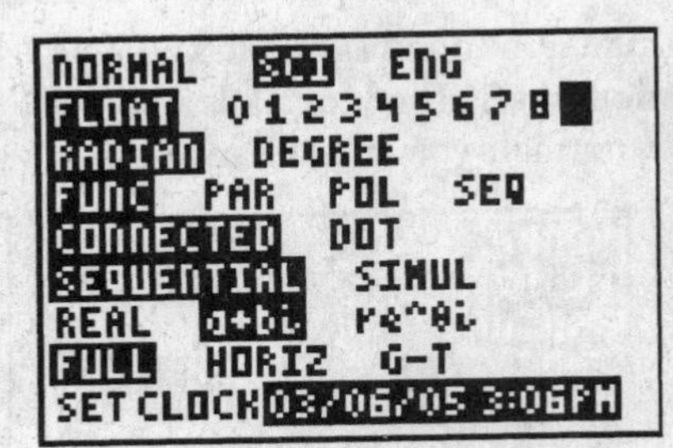

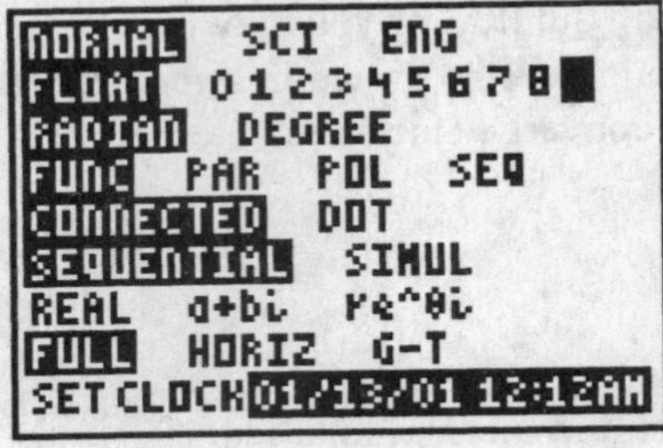

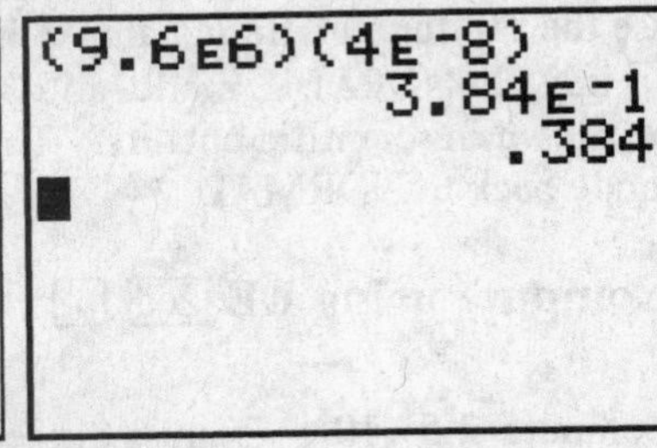

Compute $\dfrac{(2.6\times10^{-3})(4.8\times10^{-4})}{1.3\times10^{-12}}$. Express the result in scientific notation and in standard notation.

- Change the mode to SCI.
- Enter the expression using EE as shown.
- Note that it is always good practice to enclose numerators and denominators in parentheses.

- Change the mode to NORMAL.
- Press ENTER to see the result in normal notation. It is not necessary to re-enter expression.

Compute using 10^x

Compute $(9.6\times10^{6})(4\times10^{-8})$ using 10^x. Show the results in both scientific mode and in normal mode.

- Press the MODE key.
- Select SCI for scientific notation. Press ENTER.
- Press CLEAR or 2nd MODE (QUIT) to return to the Home screen.
- Enter $(9.6*10\wedge(6))(4*10\wedge(-8))$. Use 2nd LOG to enter $10\wedge($.
- Press ENTER to see the result in scientific notation.
- Change the mode back to normal to see the result in normal notation.

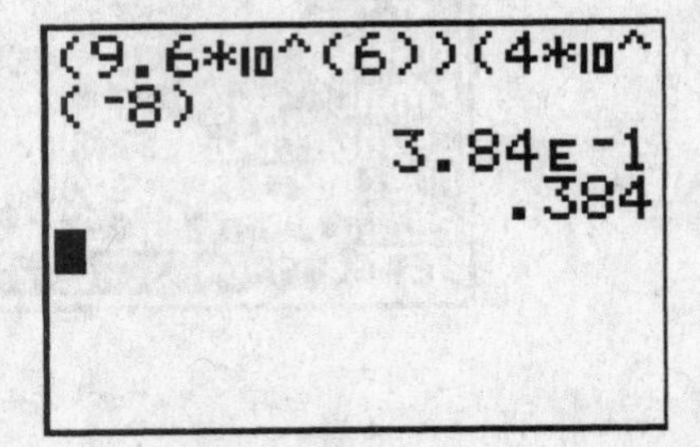

EVALUATING POLYNOMIALS

Polynomials may be evaluated on the Home screen using [STO▶], on the graph screen using CALC 1:value, or in a table.

Evaluate a Polynomial using [STO▶]

If $P(x) = -x^2 + 5x - 6$, find $P(-3)$.

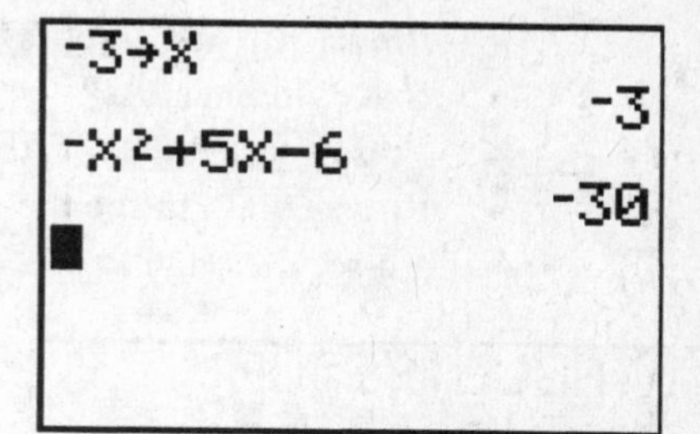

- Enter the value -3, then press the [STO▶] key and [X,T,Θ,n] to store -3 as x.
- Press [ENTER].
- Enter the polynomial using [X,T,Θ,n] for x as shown.
- Press [ENTER] to see the result.

Evaluate a Polynomial from its Graph

If $P(x) = -x^2 + 5x - 6$, find $P(-3)$.

- Graph $Y1 = -x^2 + 5x - 6$ in the standard window.
- Adjust the window until you see the point on the graph when $x = -3$.
- Press [2nd] [TRACE] (CALC).
- Select 1:value.
- Enter the value you want on the graph screen.
- Adjust the window to see the point $(-3, -30)$.

```
CALCULATE
1:value
2:zero
3:minimum
4:maximum
5:intersect
6:dy/dx
7:∫f(x)dx
```

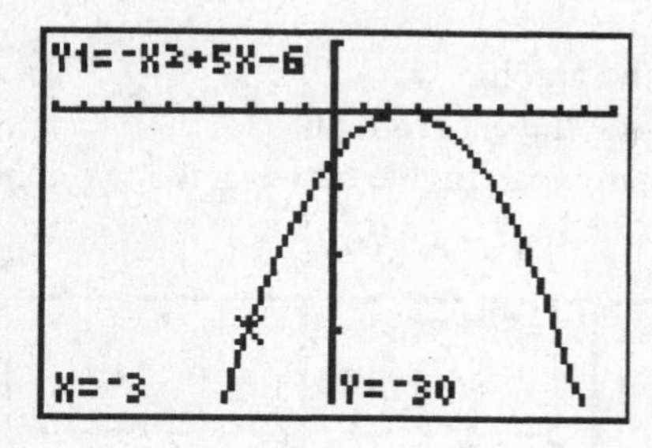

Evaluate a Polynomial from its Table

If $P(x) = -x^2 + 5x - 6$, find $P(-3)$.

- Enter $Y1 = -x^2 + 5x - 6$.
- For specific values of x, set the table to ASK.
- Press [2nd] [WINDOW] (TBL SET).
- Select Indpnt: Ask.
- Press [2nd] [GRAPH] (TABLE) to view the table. Enter the value you want, -3.
- Press [ENTER] to see the -30. You may enter more values if you wish.
- Re-set the table to AUTO to scroll through more values of the function.

```
TABLE SETUP
 TblStart=2
 ΔTbl=1
Indpnt: Auto Ask
Depend: Auto Ask
```

X	Y1
-3	-30

X=

X	Y1
-5	-56
-4	-42
-3	-30
-2	-20
-1	-12
0	-6
1	-2

X=-3

Use graphs to check the factoring of a polynomial

Check that $5x^2 + 13x - 6 = (x+3)(5x-2)$. The factoring is correct if the graphs coincide.

```
Plot1 Plot2 Plot3
\Y1=5X²+13X-6
\Y2=(X+3)(5X-2)
\Y3=
\Y4=
\Y5=
\Y6=
\Y7=
```

- Enter $Y1 = 5x^2 + 13x - 6$ and $Y2 = (x+3)(5x-2)$.
- Keep both Y1 and Y2 on.
- Press [GRAPH] to view the graphs.
- Adjust the window to see the entire graph.
- Use the up and down arrows to toggle between the two functions and see that they have the same graph.

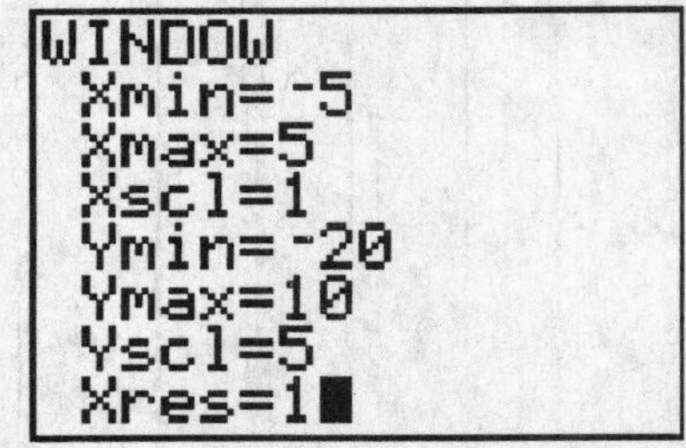

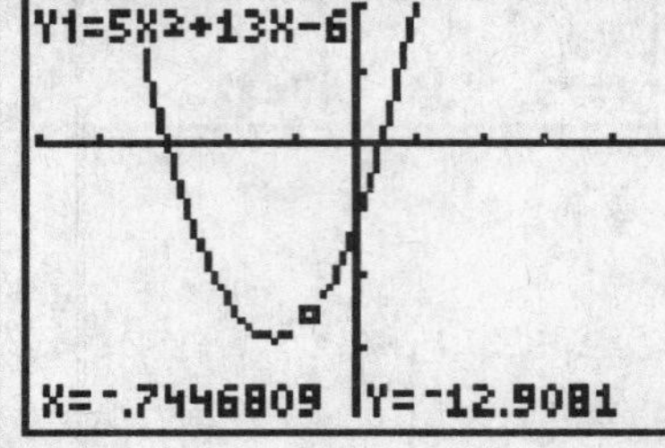

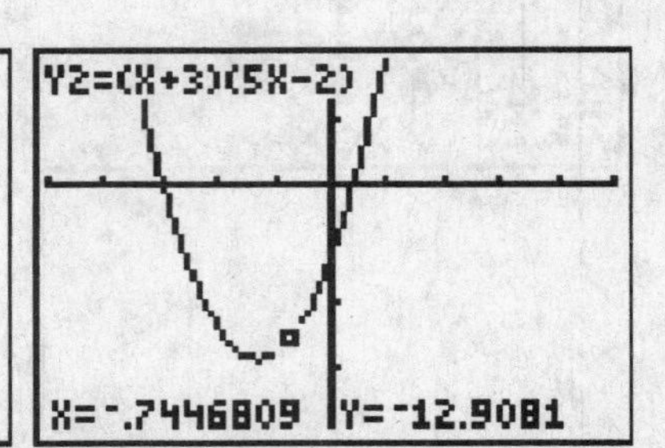

SOLVING POLYNOMIALS GRAPHICALLY

We can solve an equation graphically by solving $P(x) = 0$ and finding its x-intercepts. We can also solve an equation by setting the left and right sides of the equation as Y1 and Y2 and then finding their intersection points. These approaches were described for linear functions in Chapter 7 of this manual.

Solve Polynomials using x-intercepts

Solve $x^3 + 7x^2 = 4x + 28$.

- To solve this polynomial with x-intercepts, you must first collect all terms to the left and enter Y1 $= x^3 + 7x^2 - 4x - 28$.
- Adjust the window as shown and graph.
- Note that there are three x-intercepts.
- Use [2nd] [TRACE] (CALC) 2:zero three times to find the x-intercepts.

- Start with the leftmost x-intercept.
- Position the cursor slightly to the left of the x-intercept.
- Press [ENTER].
- Position the cursor slightly to the right of the x-intercept.
- Press [ENTER].
- Press [ENTER] or make guess about the x-intercept.
- The zero is shown at the bottom of the screen.

- Repeat the process for the remaining x-intercepts.
- Record each x-intercept.

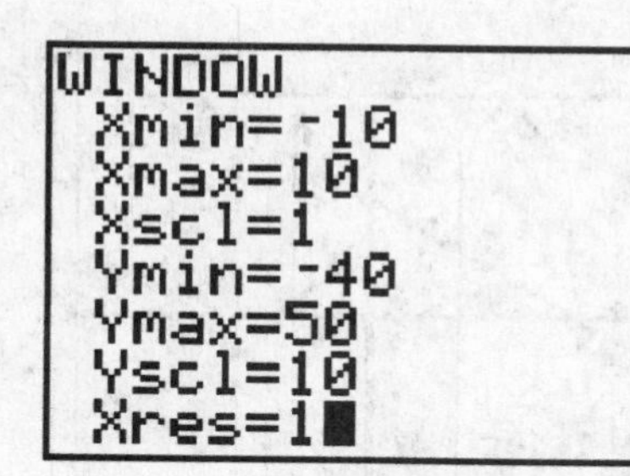

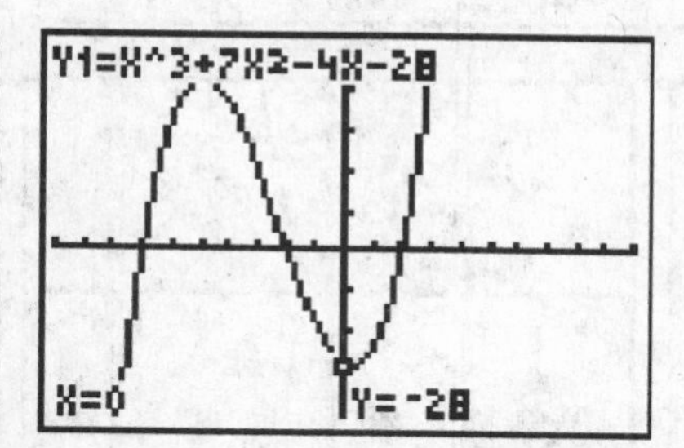

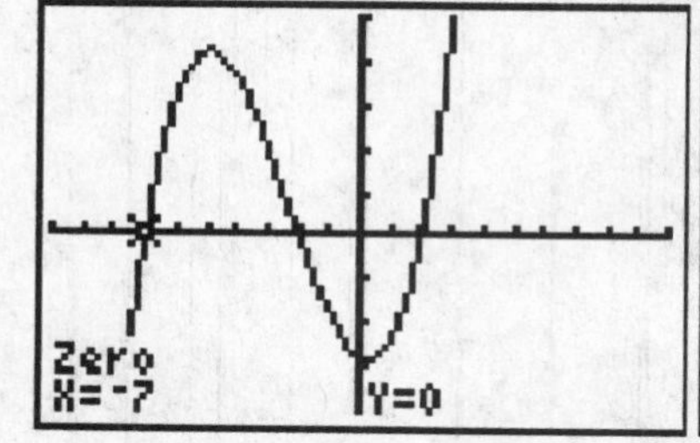

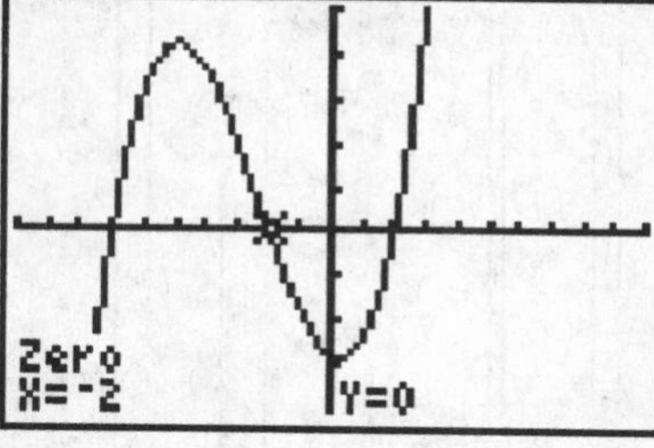

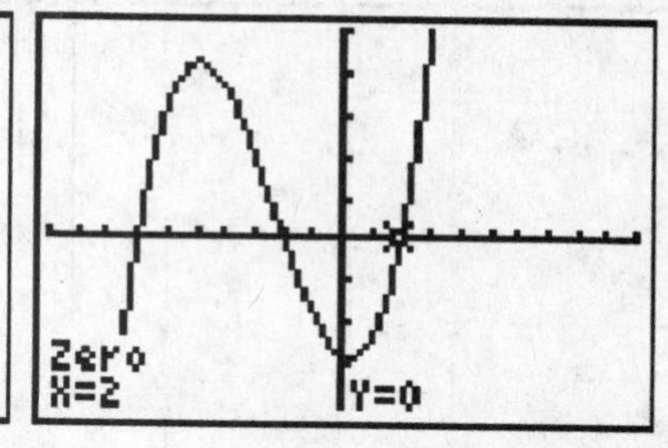

Solve polynomials using Intersection

Solve $x^3+7x^2=4x+28$.

- Enter Y1 $=x^3+7x^2$ and Y2$=4x+28$.
- Adjust the window as shown and graph.
- Note that there are three x-intercepts.
- Use [2nd] [TRACE] (CALC) 5:intersect three times to find the intersection points.

- Position the cursor near the leftmost intersection point.
- Press [ENTER] to select the first curve.
- Press [ENTER] again to select the second curve.
- Press [ENTER] to see the Intersection point at the bottom of the screen.

- Repeat the procedure for the remaining intersection points.
- Record each intersection point.

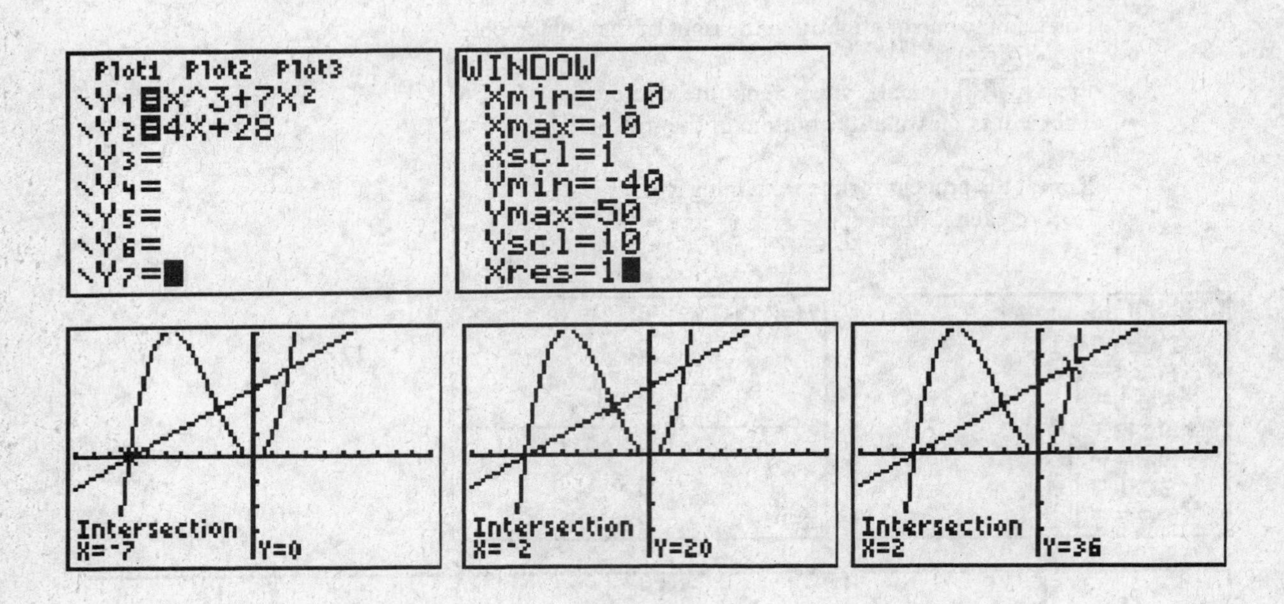

CHAPTER 11

RATIONAL EXPRESSIONS AND FUNCTIONS

EVALUATING RATIONAL EXPRESSIONS

Rational expressions are made up of the form $\frac{P(x)}{Q(x)}$, where $P(x)$ and $Q(x)$ are polynomials. Since rational expressions are fractions, the denominator, $Q(x) \neq 0$.

The calculator will give you an error message if you try to evaluate a rational expression at a value that makes the denominator zero. Also, you must enclose the numerator and denominator in parentheses when they have more than one term.

Evaluate $\frac{7x+2}{x-3}$ at $x=-1$, $x=3$, $x=5$. Convert the result to a fraction.

- Store −1 in x by using [X,T,Θ,n].
- Press [(-)] [1] [STO▸] [X,T,Θ,n].
- Enter the expression as $(7x+2)/(x-3)$.
- Press [ENTER] to see the result, 1.25.

- To convert the answer to a fraction, press [MATH] select 1:➔FRAC.
- Press [ENTER] to see the result, $\frac{5}{4}$.
- Note that you can do the computation and the conversion to a fraction on one line.

- Store 3 to x.
- Use [2nd] [ENTER] until you get to the rational expression.
- Press [ENTER].
- This time you get an error message that you divided by 0.
- Press [1] to Quit or [2] to go to the problem.

- Store 5 to x.
- Use [2nd] [ENTER] until you get to the rational expression.
- Press [ENTER] and record the result, $\frac{37}{2}$.
- Convert to a fraction on the same line as shown.

Guidelines for Graphing Rational Functions

Rational functions must be graphed with care. They are often made up of two or more terms in the numerator and the denominator. Also, fractions may not have a denominator of zero, so there will be values of x that will not be used on the graph. To help with these issues,

- Enclose both the numerator and denominator in parentheses when they contain more than one term.
- Graph the rational function with dots instead of connected lines.

Two Ways to Graph with Dots

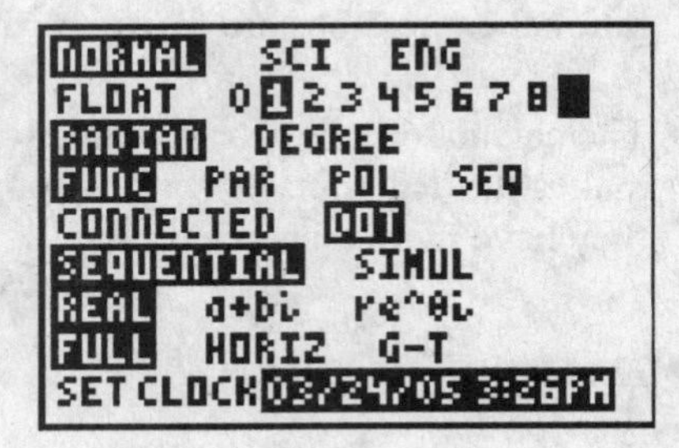

1. Set the MODE on the calculator to DOT. When you do this, all the graphs will be drawn with dots.
   - Press the [MODE] key.
   - Use the arrow keys to go to the word DOT.
   - Press [ENTER] to set the graphing mode to DOT.
   - All graphs will be drawn with dots.

2. If the calculator is in Connected mode, you may graph an individual function with dots from the Y= screen.
   - Press the [MODE] key.
   - Use the arrow keys to go to the word CONNECTED.
   - Press [ENTER].
   - Use the left arrow key to place the cursor to the left of the Y= function you wish to use.
   - Press [ENTER] until you see three dots. That equation alone will be graphed in DOT mode.
   - The graph is drawn in the window shown.
   - Press [GRAPH], then [TRACE]. Only points that are on the function are shown.

   - Press [3] [ENTER]. Note that there is no y-value shown. This is because this function is not defined at $x = 3$.
   - Move the cursor to another point and you will see both the x and the y-values for that point.
   - Note that if you clear the equation, the dot setting disappears.

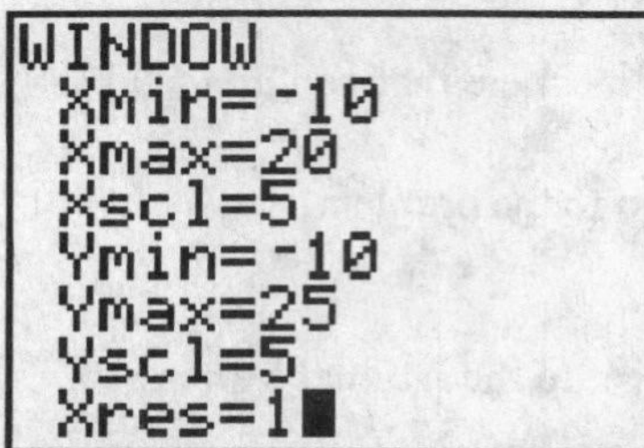

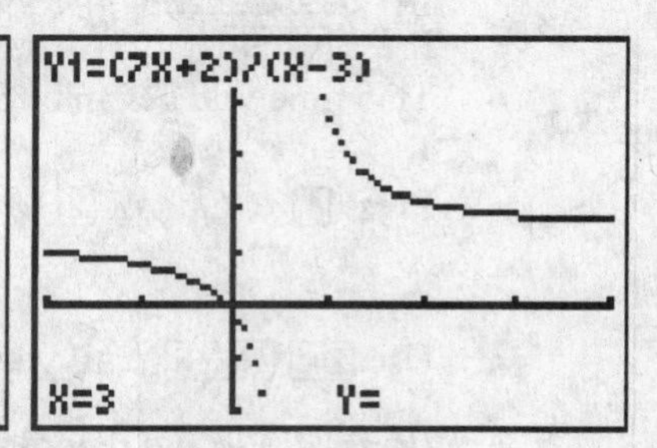

Solve Rational Equations Graphically using Intersection

Solve $\frac{3}{x}-\frac{x+21}{3x}=\frac{5}{3}$.

- Enter the left side of the equation as Y1= and the right side as Y2=.
- Look at the graph of each side separately to get a better understanding of the graphs.
- Note that the denominator of the second term *MUST* be in parentheses to graph correctly.
- Graph Y1= in the window shown below.
- Note that there is no y-value is shown when $x=0$ because the function is undefined there.

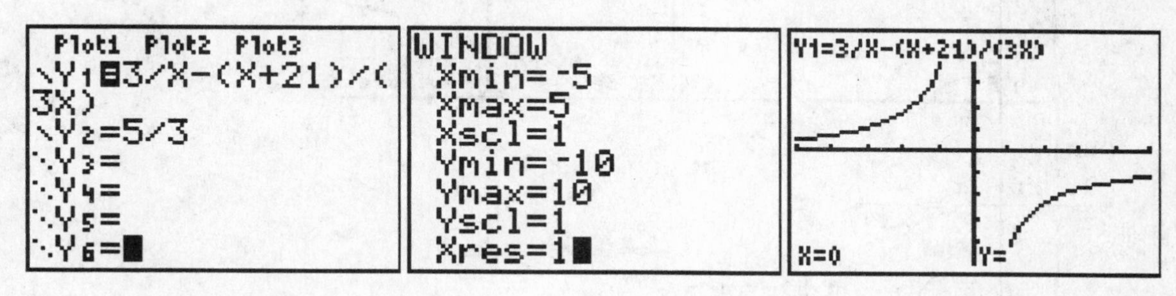

- Turn off Y1= and turn on Y2=.
- To turn a function on or off, move the cursor over the = sign and press [ENTER].
- Note that $y=\frac{5}{3}$ is a horizontal line.

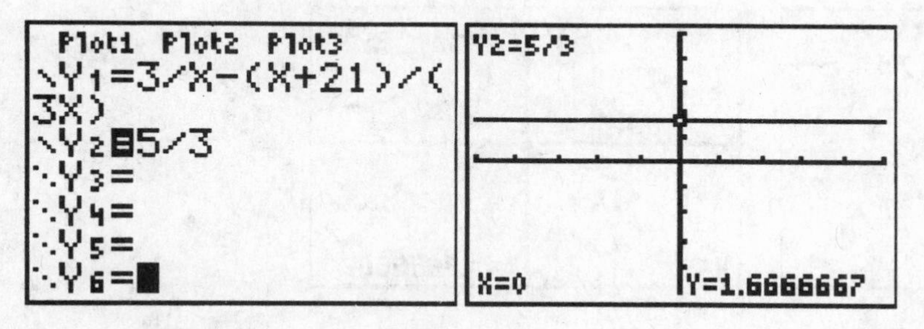

- Turn Y1= back on.
- Press [GRAPH] to graph both functions.
- Use [2nd] [TRACE] (CALC) to identify the intersection point.
- Select 5:intersect.

- Move the cursor near the intersection.
- Press [ENTER] to select the first curve.
- Press [ENTER] to select the second curve.
- Press [ENTER] to see the coordinates of the intersection point at $(-2, 1.6666667)$ or $(-2, \frac{5}{3})$.

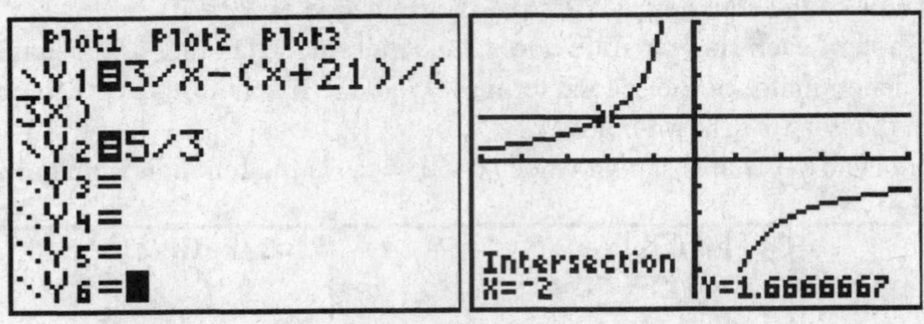

Solve $\frac{2x}{x-3}+\frac{6-2x}{x^2-9}=\frac{x}{x-3}$.

- Enter the left side of the equation as Y1=
- Enter the right side of the equation as Y2=.
- I recommend that you look at the graph of each side separately to get a better understanding of the graphs.
- Graph Y1= in the standard window.

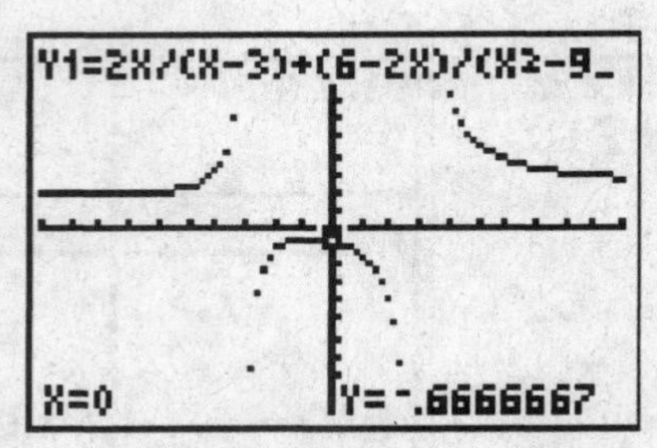

- Turn off Y1=.
- Turn on Y2=.
- To do this, move the cursor over the = sign and press [ENTER].
- Graph Y2= in the standard window.

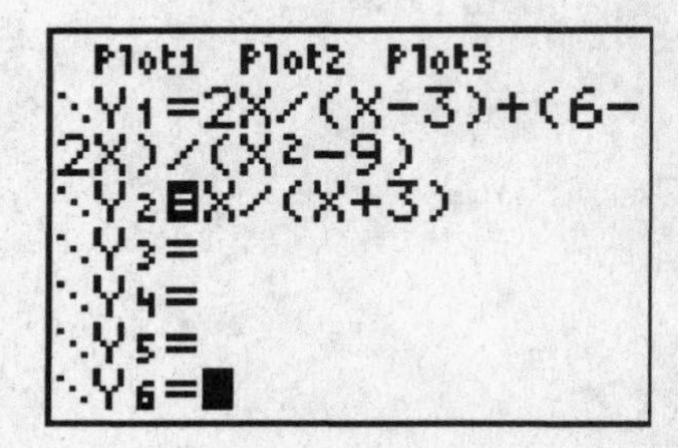

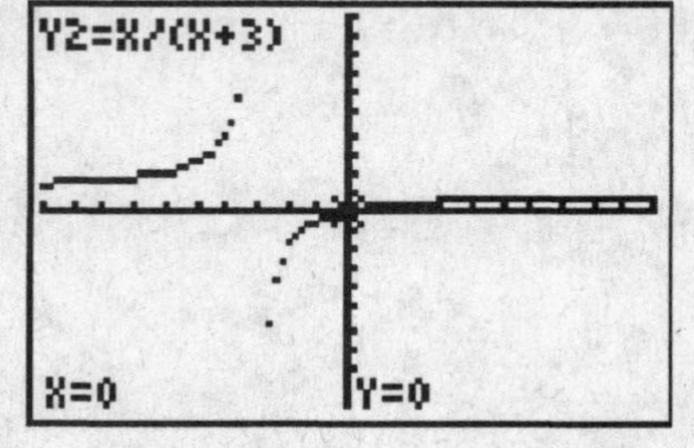

- Now turn Y1= back on and graph both functions in the standard window.
- Adjust the window as shown to get a better look at the intersection points.

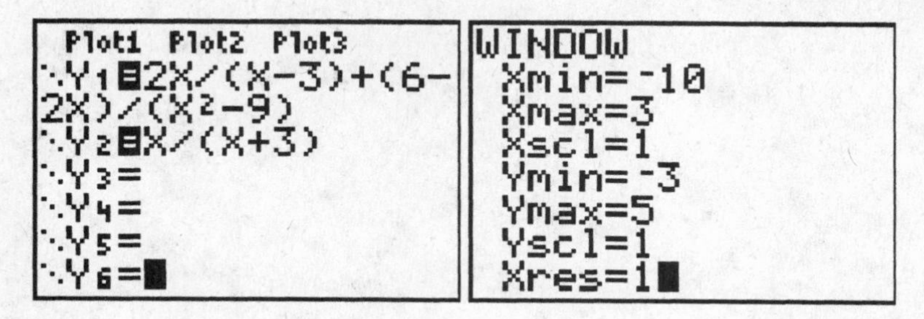

- Use [2nd] [TRACE] (CALC) twice to identify the two intersection points.
- Select 5:intersect.
- Move the cursor near the intersection point on the left side of the screen.
- Press [ENTER] three times to see the intersection at $(-6, 2)$.
- Repeat the above steps to identify the second intersection point at $(-1, -.5)$.

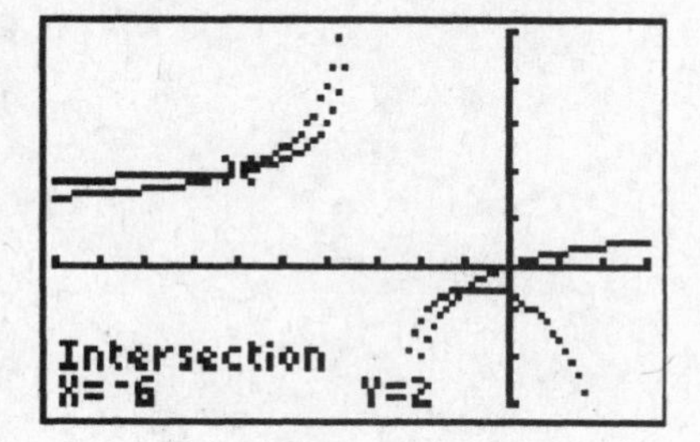

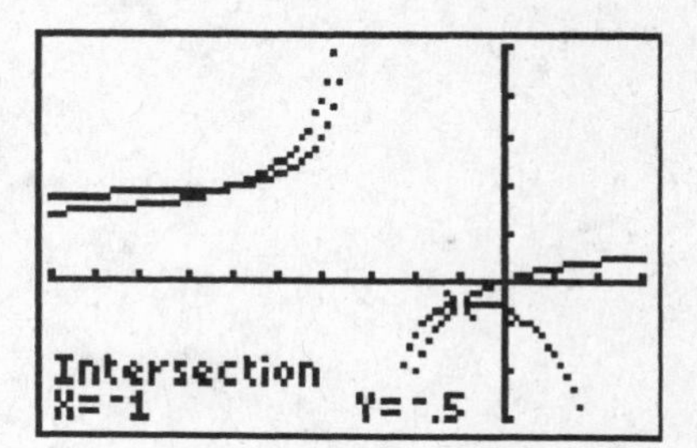

CHAPTER 12

LINEAR INEQUALITIES

LINEAR INEQUALITIES IN ONE VARIABLE

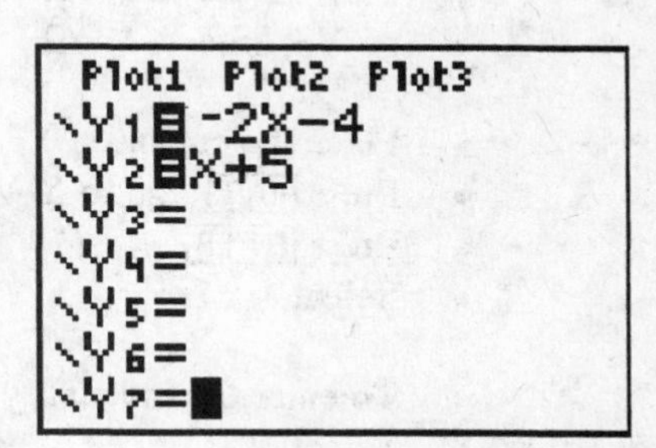

Solve a Linear Inequality in One Variable by Graphing

Solve the inequality $-2x-4>x+5$ by graphing.

Use Intersection to solve $-2x-4>x+5$

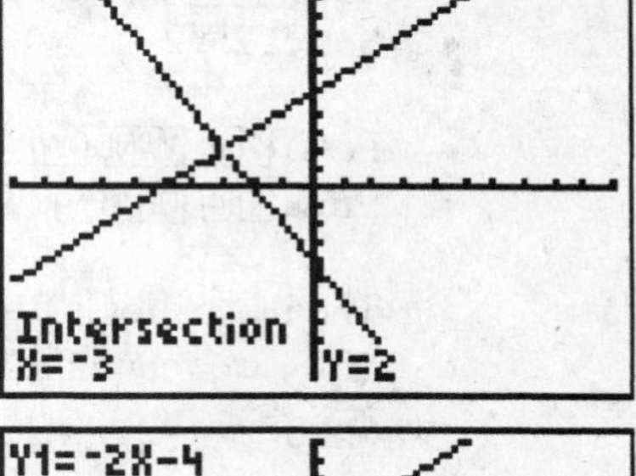

- Turn off or clear any equations or plots.
- Enter the left side as Y1=$-2x-4$
- Enter the right side as Y2=$x+5$.
- Set the standard window.
- Use [2nd] [TRACE] (CALC) to find the intersection of Y1=$-2x-4$ and Y2=$x+5$.
- Identify the Intersection as the point $(-3,2)$.
- Press [TRACE].
- Move the cursor up or down to locate which graph is Y1=$-2x-4$, and which graph is Y2=$x+5$.
- Move along Y1 until you see the Y1 line above the Y2 line.
- Notice that Y1 is above Y2 to the left of the intersection point $(-3,2)$, so Y1 > Y2 when $x<-3$ over the interval $(-\infty,-3)$.

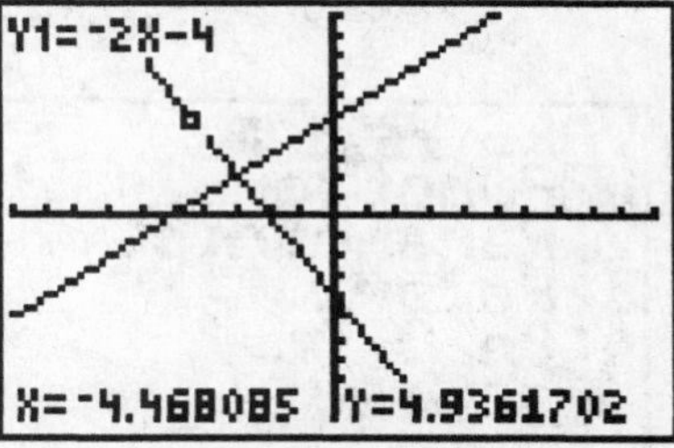

Use a Table to solve $-2x-4>x+5$.

- Enter Y1=$-2x-4$; enter the right side as Y2=$x+5$ as described above.
- Press [2nd] [WINDOW] (TBLSET) to set the table. Start at -10 and use $\Delta Tbl=1$. Set both Indpnt and Depend to AUTO.
- Press [2nd] [GRAPH] to view the table.
- Notice that the values of Y1 are greater than the values of Y2 to the left of -3.
- Scroll down through the x-column until you see the values of Y1 become smaller than the values of Y2.
- Therefore, Y1 > Y2 when $x<-3$ over the interval $(-\infty,-3)$.

X	Y1	Y2
-6	8	-1
-5	6	0
-4	4	1
-3	2	2
-2	0	3
-1	-2	4
0	-4	5

X= -3

Use the Test menu and a Table to solve $-2x-4>x+5$

- Set the Standard window using ZOOM 6.

- Turn off Y1 and Y2.
- At Y3, enter Y1>Y2.

- Place the cursor on Y3.
- Press [VARS], go to Y-VARS and select 1:Function.
- Press [ENTER].
- Select 1:Y1.

- To enter the inequality symbol press [2nd] [MATH] (TEST).
- Select 3:>.
- Use [VARS] again to enter Y2. Go to Y-VARS and select 1:Function.
- Select 2:Y2

- Press [2nd] [WINDOW] (TBLSET) to set the table as shown.
- Press [2nd] [GRAPH] (TABLE) to view the table of Y3.

- The inequality is true wherever there is a 1, and false wherever there is a 0.
- Since we found the intersection of the lines to be at the point $(-3,2)$ and there are 1's to the left of $x=-3$, we conclude that Y1 > Y2 is true when $x<-3$ over the interval $(-\infty,-3)$.

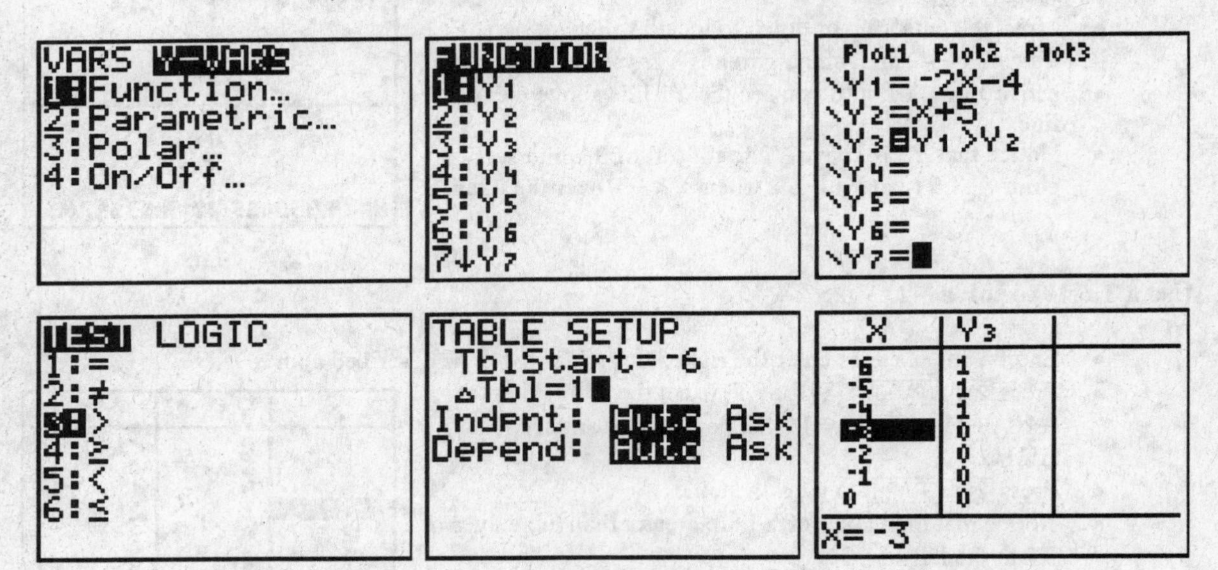

Graph Compound Linear Inequalities in One Variable

Use TEST and TEST LOGIC to solve the intersection (and) or union (or) of compound inequalities. To do this, first enter each side of each inequality and only look at the graph and table of the final relationship.

Solve the Compound Inequality with OR:

$2x-3<7$ or $35-4x\leq 3$.

Enter all Relationships in Y=

- Change the MODE to DOT.
- Enter the inequality symbols using 2nd MATH (TEST).
- Enter each side of $2x-3<7$ as Y1 and Y2.
- Enter Y3 = Y1 < Y2 .

- Use VARS Y-VARS 1:Function to enter the function names.
- Use 2nd MATH (TEST) to enter the inequalities.

- Enter $35-4x\leq 3$ as Y4 and Y5.
- Enter Y6 =Y4 ≤ Y5.

- Enter Y7 = Y3 or Y6.
- Use 2nd MATH LOGIC to enter the word OR.
- Only turn on Y7. Press GRAPH.

- Use TRACE to move along the union.
- Note that you do not necessarily get the exact values when you trace along Y7.

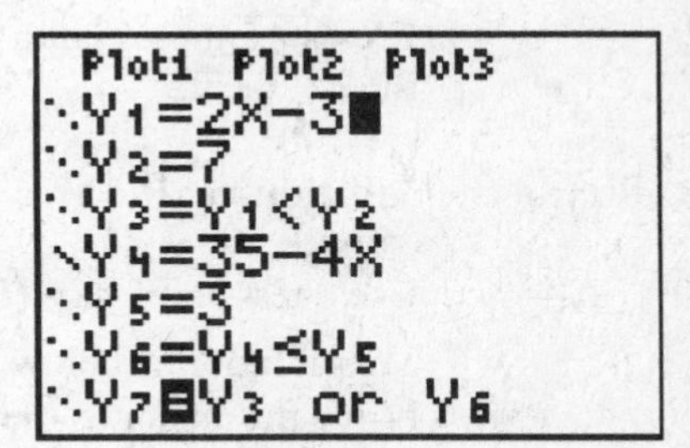

- View the table of Y7. Scroll through the table until you see both 1's and 0's. The inequality is true wherever there is a 1, and false wherever there is a 0 .

- The solution is the set $\{x|x<5 \text{ or } x\geq 8\}$ $(-\infty,5)\cup[8,\infty)$.

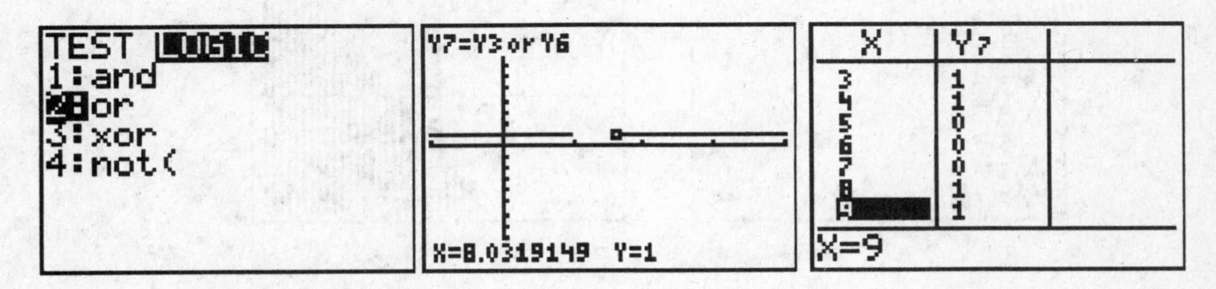

LINEAR INEQUALITIES IN TWO VARIABLES

Solve a Linear Inequality in Two Variables by Graphing using Shading.

Solve the inequality $6x-3y>15$ by graphing $y<2x-5$.

Enter the function

- Turn off or clear any equations or plots.
- Solve the equation for y and enter it as Y1.
- Enter Y1=$2x-5$.
- You are solving the equivalent inequality, $y<2x-5$.

Select solid or dotted line using MODE

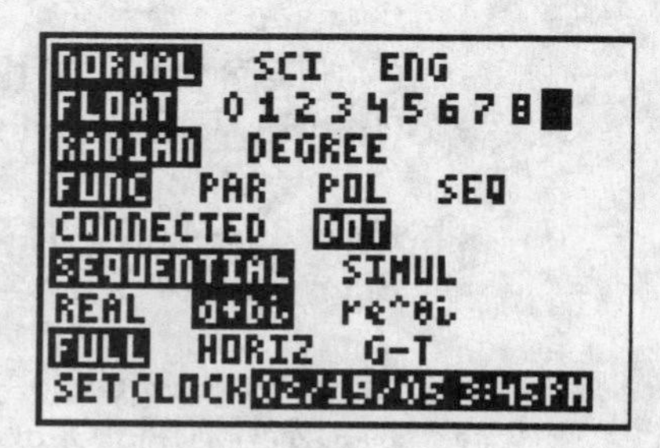

- Since this is a strict inequality, graph the line as dotted. To do this, press the MODE key.
- Cursor down to the line that reads Connected Dot.
- Select Dot.
- Press CLEAR to return to the Home screen or Y= to go to the Y= menu.

Shade the solution region

- Use a test point, say $(0,0)$, to decide whether the solution is above or below the line.
- This solution is below the line because $0<-5$ is not true.
- Move the cursor to the very left of the ⋱Y= line. The ⋱ will be flashing.
- Press ENTER until you see ◣ symbol. This means that you want the graph to be shaded below the line.
- Move the cursor away from the symbol to select it.
- Press GRAPH to see the shaded graph.

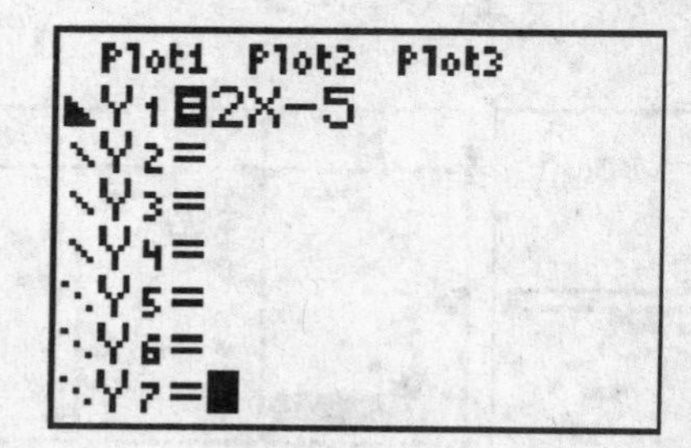

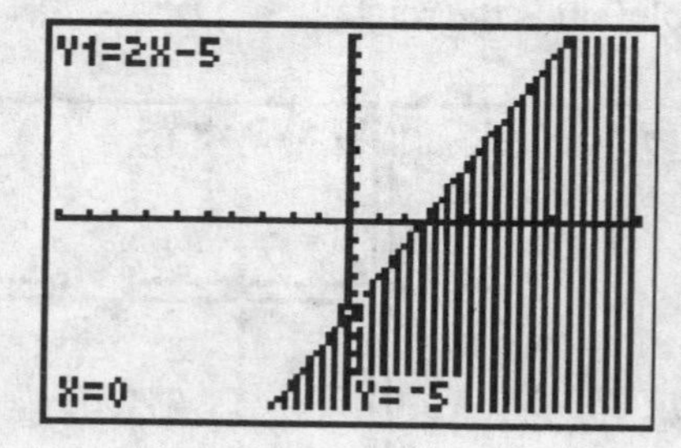

Graph a System of Linear Inequalities in Two Variables using Shading

Graph the solution set: $\begin{matrix} x-y<1 \\ 2x+3y \geq 12 \end{matrix}$

- Solve each inequality for y.
- Enter Y1 $> x-1$ and Y2 $\geq 4-(2/3)x$.
- Move the cursor to the very left of the ⋱Y1= line over the ⋱. The ⋱ will be flashing.
- Press [ENTER] until you see ◥ symbol. This means that you want the graph of Y1 to be shaded above the line.
- Move the cursor away from the symbol to select it.

- Move the cursor to the very left of the ⋱Y2= line over the ⋱. The ⋱ will be flashing.
- Press [ENTER] until you see ◥ symbol. This means that you want the graph of Y2 to be shaded above the line.
- Move the cursor away from the symbol to select it.

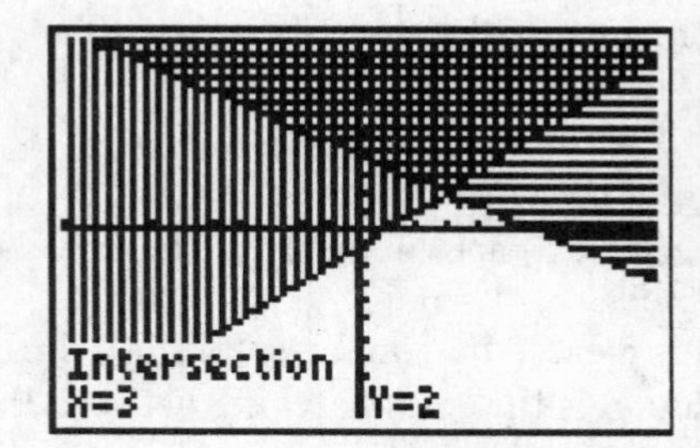

- Press [GRAPH] to see the graph of the two inequalities.
- The solution set is the area that includes the cross-hatched shading at the top of the screen.

- Use CALC to find the intersection point.
- This system has only one intersection point, $(3, 2)$, which is not in the solution set of the system because one of the lines is dotted.

Graph a System Linear Inequalities with Vertical and/or Horizontal lines using Shading

Graph the solution set: $x - y < 2$, $-2 \le x < 4$, $y < 3$

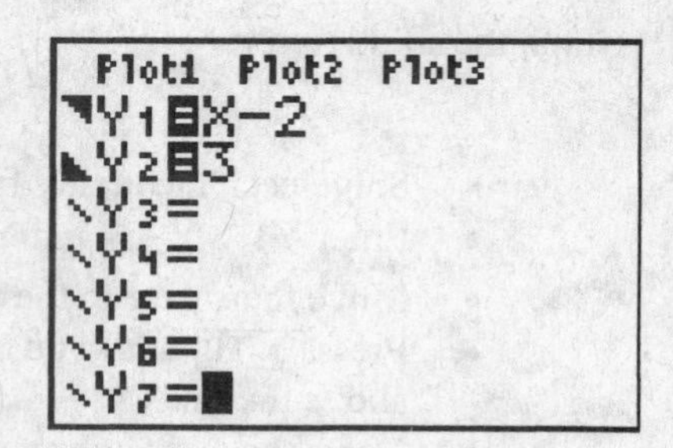

- Solve the first inequality for y, $y > x - 2$.
- Enter $x - 2$ as Y1.
- Select the shading to be above the line.
- Enter the horizontal line 3 as Y2.
- Select the shading to be below the line.
- Next, the vertical lines $x = -2$ and $x = 4$ must be graphed from the Home screen using the DRAW menu.
- From the Home screen press [2nd] [PRGM] (DRAW) to access the Draw menu.
- In the DRAW submenu scroll down to 4:Vertical or press the [4] key. This puts the word Vertical on the Home screen.
- Press a number to draw a vertical line. For example, the number -2 will graph the line $x = -2$. Repeat for $x = 4$.
- Press [ENTER] to view the graph.
- Note that [TRACE] does not work. You may use the arrow keys to move the cursor around the graph screen.
- You must clear a drawing by selecting 1:ClrDraw in the DRAW menu.
- The solution is the cross-hatching between the vertical lines $x = -2$ and $x = 4$.
- You can identify the corners of the solution by using CALC 1:value to find the value of Y1 when $x = -2$ and $x = 4$, and by finding the value of Y2 when $x = -2$ and $x = 4$.
- The four corners are $(-2,-4)$, $(4,2)$, $(-2,3)$, $(4,3)$.
- Note that the intersection of Y1 and Y2 is $(5,3)$, which is outside of the region.

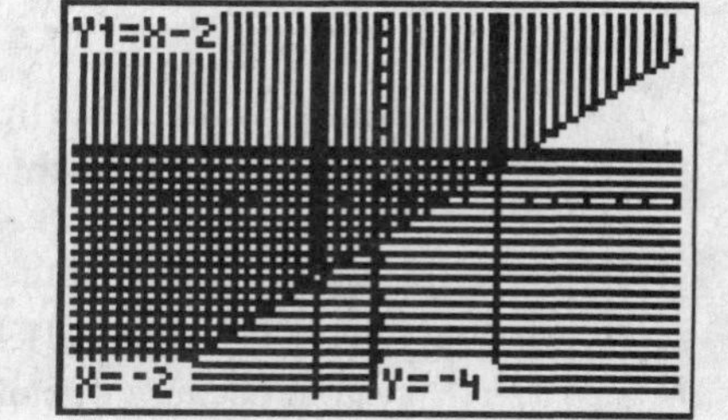

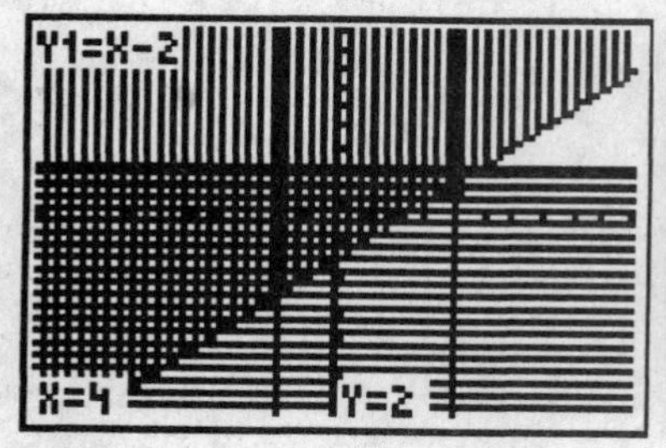

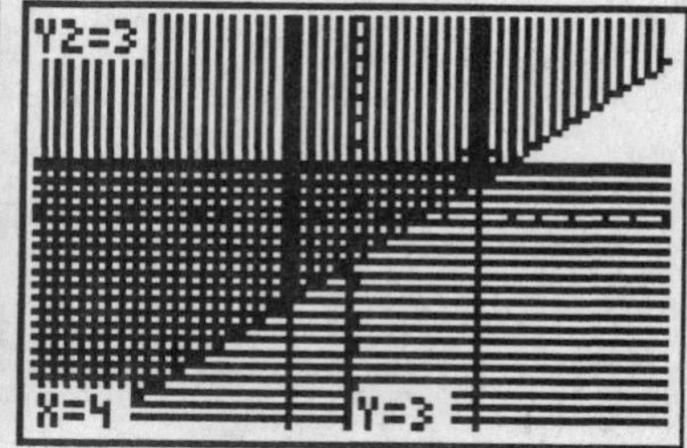

Graph a Linear Inequality using Inequalz Apps

The TI-84 Plus Silver Graphing Calculator has many applications built into the calculator. Press the APPS key to see all the choices. Inequalities are more easily graphed using the Inequalz App.

This Inequalz application can be found on the Texas Instruments website for the TI-83 Plus graphing calculator.

Graph $y < x - 5$.

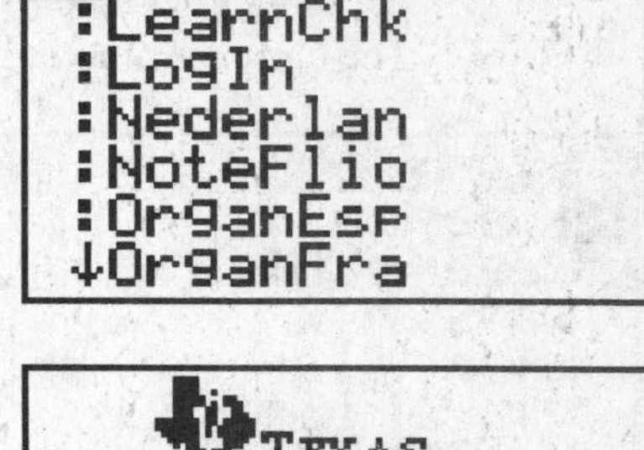

- Press the APPS key. It is in the third row of keys next to the [MATH] key.
- Press the green [ALPHA] key, then the [x^2] key to jump to the I's. Select Inequalz.
- You see a Texas Instruments INEQUALITY GRAPHING SCREEN.
- Press any key to continue.

- You are taken to the Y= screen.
- Notice that there are inequality symbols at the bottom of the screen.
- Place the cursor over the = sign and press the green [ALPHA] key and the key below the inequality you want, < , the [ZOOM] key in this example.
- Notice that there is now a < symbol at Y1, and that the shading will be below the line.

- Set the Standard Window.
- Press the [GRAPH] key to see the graph.
- Press [TRACE] to move along the dotted line.
- The bottom of the screen shows three choices, SHADES, PoI-Trace, and ? These options are for systems of two or more inequalities.
- You must use the green [ALPHA] key to use these options.

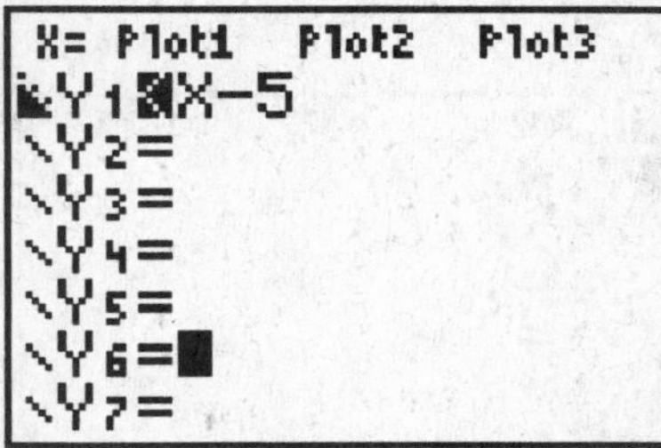

Shades PoI-Trace ?

Graph a System Linear Inequalities on the TI-84 Plus Silver using Inequalz Apps

Graph the solution set: $\begin{array}{l} x-y<1 \\ 2x+3y\geq 12 \end{array}$

- Press the APPS key. It is in the third row of keys next to the [MATH] key.
- Press the green [ALPHA] key, then the [x^2] key to jump to the I's.
- Select Inequalz. You are brought to the opening screen of Inequality Graphing.
- Press any key to begin. You are brought to a new Y= screen with inequality symbols at the bottom of the screen.
- The cursor is blinking over the = sign.

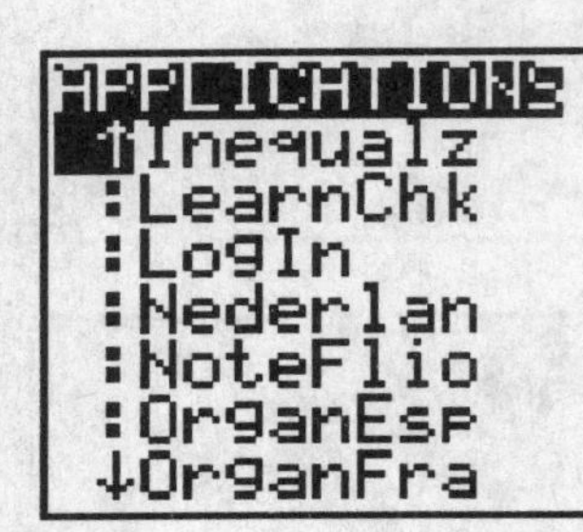

- Enter $\mathrm{Y1} > x-1$ and $\mathrm{Y2} \geq 4-(2/3)x$.

- To enter the inequality symbols, place the cursor over the = sign, use the green [ALPHA] key and the key directly under the inequality you want.

- To select >, press [ALPHA] [TRACE].
- To select ≥, press [ALPHA] [GRAPH].

- Note that the inequality symbols are only shown when the cursor is over the = sign.
- Note also that the shading symbol to the left of Y1 and Y2 are shown as shading above the line.

- Press [GRAPH] to see the graph of the two inequalities.

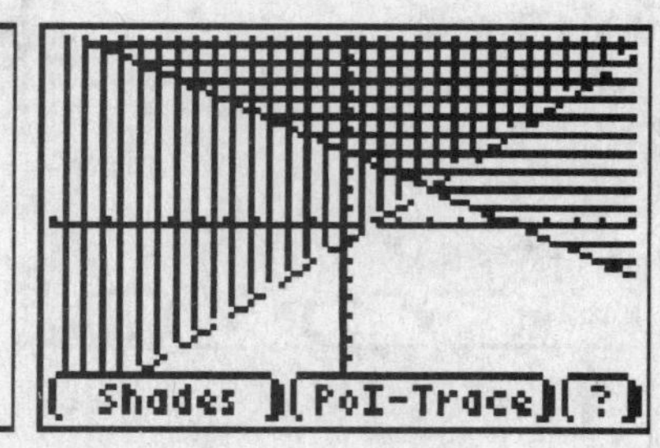

Shade the intersection of the solutions only

The graph screen has three items at the bottom of the screen, Shades, PoI-Trace and ?.

- We wish to shade the intersection of the two inequalities because we are solving a system of linear inequalities.
- Press the green [ALPHA] key, then either the [Y=] or [WINDOW] key to select SHADES.
- Move the cursor to 1:Ineq Intersection.
- Press [ENTER] to select Ineq Intersection.

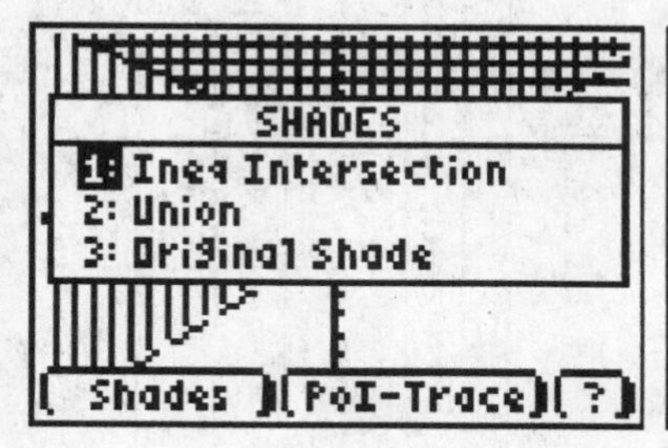

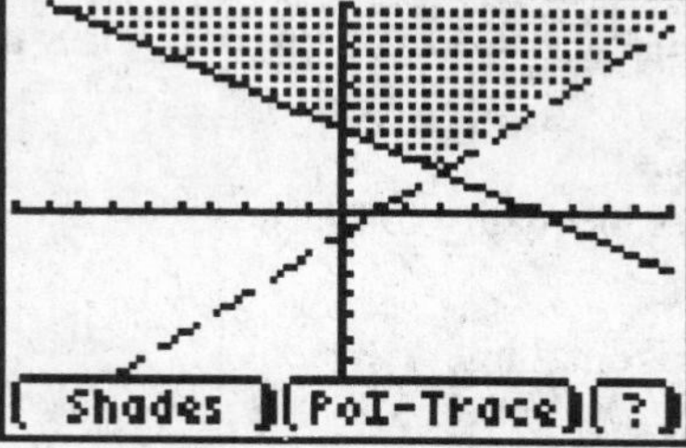

Find the point(s) of intersection or corners of the solution

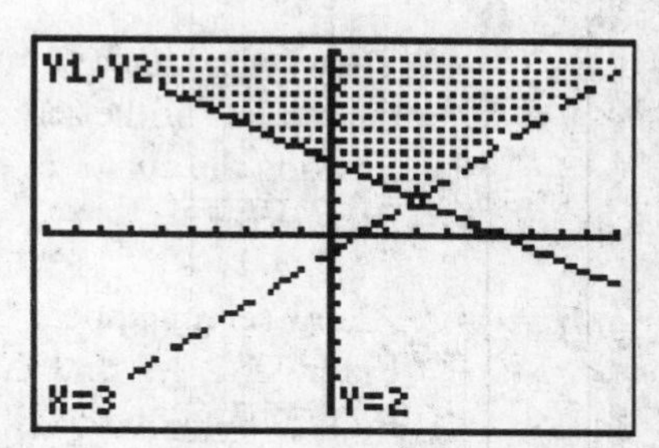

- Use the second item, PoI or Points of Intersection, to identify the corners of the solution.
- Press [ALPHA] [ZOOM] or [ALPHA] [TRACE] to select PoI.
- PoI-Trace locates any intersection points of the linear system.
- This system has only one intersection point, $(3, 2)$, which is not in the solution set of the system because one of the lines is dotted.
- When the system has more than one corner point, use the arrow keys to move from corner to corner.

Graph a System Linear Inequalities with Vertical and Horizontal lines on the TI-84 Plus Silver using Inequalz Apps

$$x - y < 2$$

Graph the solution set: $-2 \le x < 4$

$$y < 3$$

- Press the APPS key. It is in the third row of keys next to the [MATH] key.
- Press the green [ALPHA] key, then the [x^2] key to jump to the I's.
- Select Inequalz.
- Solve the first inequality for y, $y > x - 2$.
- Enter Y1 $> x - 2$.
- Enter the horizontal line Y2 < 3.

Graph the vertical lines

- The vertical lines $x = -2$ and $x = 4$ can be graphed from the graph screen by going up to the top left of the screen over X=.
- Position the cursor over X=.
- Press [ENTER].

- A new screen appears for entering X= equations.
- Enter X1 ≥ -2 and X2 < 4 as shown.
- Press [GRAPH] to see all the graphs of the inequalities.

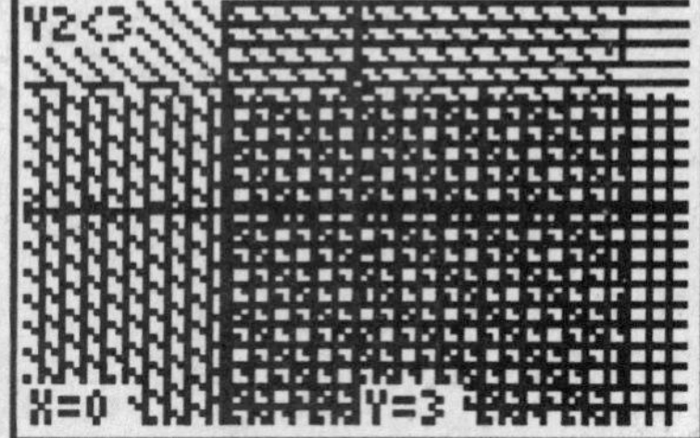

- The solution is the cross-hatching of all the lines.
- Notice that this is very difficult to decipher.

Use Shades to shade the intersection of all the inequalities

- You see the word Shades at the bottom of the screen.
- Press [ALPHA] [Y=] or [ALPHA] [WINDOW] to see the Shades menu.
- Move the cursor up to 1:Ineq Intersection.
- Press [ENTER] to select Ineq Intersection.

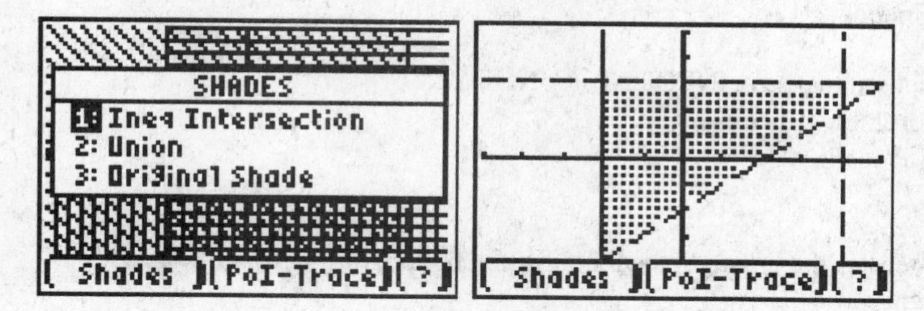

- Use PoI-Trace to identify the corners.
- Press [ALPHA] [ZOOM] or [ALPHA] [TRACE] to jump to one of the intersection or corner points of the solution.
- Use the arrow keys to identify the remaining corner points.
- Note that it may take several tries to reach all the corners.

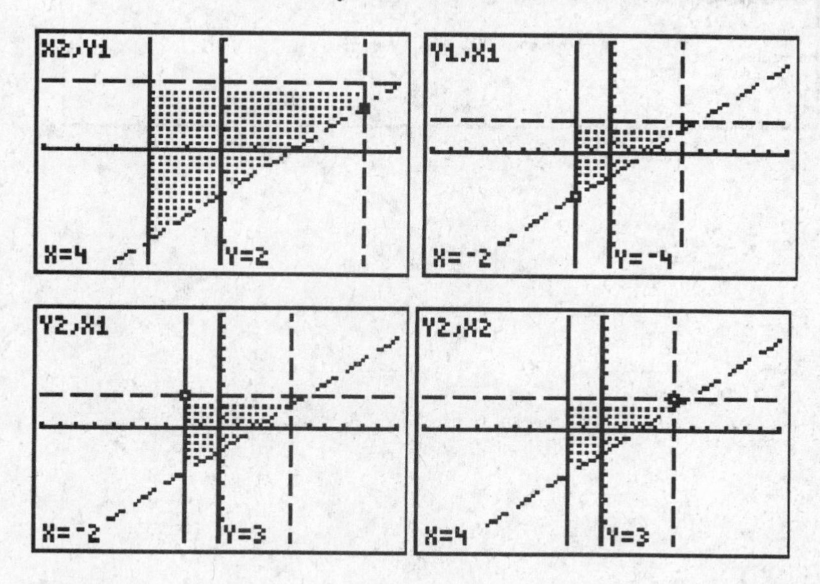

CAUTION: The Points of Intersection feature will also identify the intersection of the two lines, identified as as $Y1 = x - 2$ *and* $Y2 = 3$ *since it is also an intersection. However, we are only interested in the intersections of the lines which form the solution to the inequalities. The intersection of Y1 and Y2* $(5,3)$ *is not a solution of the system of inequalities because it lies outside the region of solutions for the system of inequalities.*

To Quit INEQUALZ

- To quit Inequalz, you must again locate Inequalz with the APPS key.
- Press APPS.
- Press [ALPHA] [x^2] to jump to the I's.
- Position the cursor at Inequalz.
- Press [ENTER] to select Inequalz.

- You are brought to a screen that says INEQUAL RUNNING.
- Position the cursor over 2: Quit Inequal.
- Press [ENTER] to leave Inequalz.

- The Y= screen no longer has the inequality symbols at the bottom of the screen when the cursor is over the = sign.

```
APPLICATIONS
↑Inequalz
 :LearnChk
 :LogIn
 :Nederlan
 :NoteFlio
 :OrganEsp
↓OrganFra
```

```
INEQUAL RUNNING
1:Continue
2:Quit Inequal
3:About
```

```
Plot1 Plot2 Plot3
\Y1=
\Y2=
\Y3=
\Y4=
\Y5=
\Y6=
\Y7=
```

CHAPTER 13

RADICALS AND RADICAL FUNCTIONS

EVALUATING RADICAL EXPRESSIONS

Radical expressions are made up of the form $\sqrt[x]{n}$, where x is called the index and tells what kind of root you are taking, such as square root, cube root, fourth root, etc., and n is called the radicand. In this course we are mostly interested in real roots of radicals, so set the MODE to Real.

However, the TI-83 Plus and the TI-84 Plus will evaluate complex roots if the MODE is set to $a+bi$. When the calculator mode is set to Real, the calculator will give you an error message if you try to evaluate a radical expression at a value that gives a complex number, namely when the index is even and the radicand is negative.

Evaluate Square Roots

Use the square root key to evaluate $\sqrt{16}$, $-\sqrt{64}$, $\sqrt{-9}$. The square root key is a second function of the [x^2] key. The radicand of a square root must be positive to give a real number.

- Set the MODE to Real.
- To evaluate $\sqrt{16}$, enter the square root symbol first.
- Press [2nd] [x^2] ($\sqrt{\ }$).
- Enter the value, 16.
- Press [ENTER] to see the result, 4.

- To evaluate $-\sqrt{64}$, enter the negative first, then the square root symbol.
- Enter the value, 64.
- Press [ENTER] to see the result, -8.

- To evaluate $\sqrt{-9}$, enter the square root symbol first.
- Enter the value, -9.
- Press [ENTER] to see the result. You get an error message as shown.

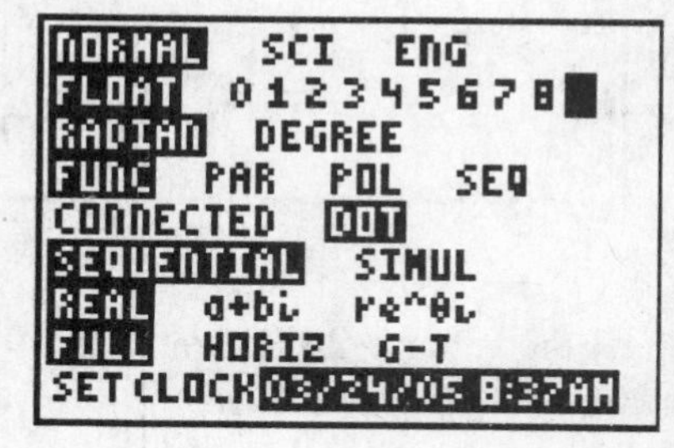

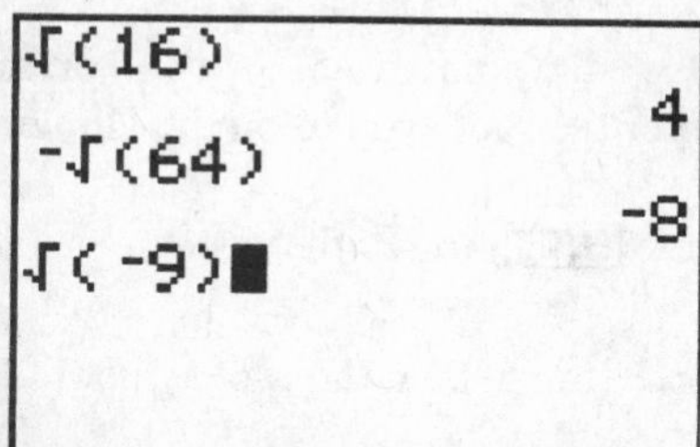

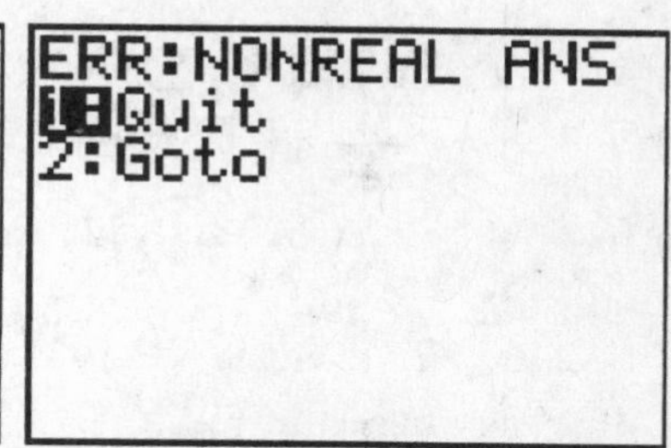

Evaluate Cube Roots

Cube roots can be evaluated using the $\sqrt[3]{\ }$ in the MATH menu. The radicands of cube roots may be either positive or negative to give a real number.

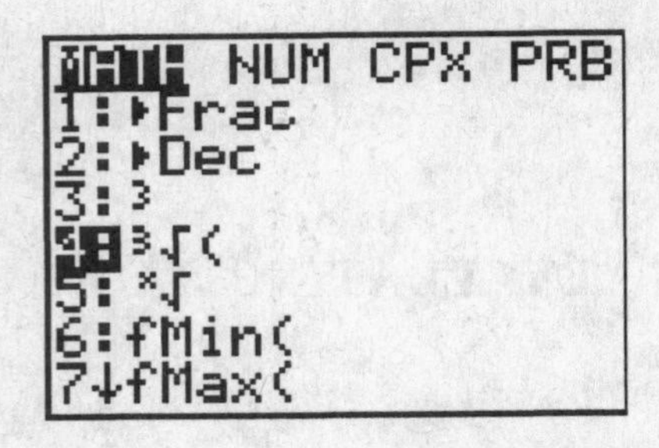

Evaluate $\sqrt[3]{8}$, $-\sqrt[3]{27}$, $\sqrt[3]{-64}$.

- To evaluate $\sqrt[3]{8}$, press the [MATH] key.
- Move the cursor to 4: $\sqrt[3]{\ }$ (and press [ENTER], or press [4] to put the cube root symbol on the Home screen.
- Enter the value, 8.
- Press [ENTER] to see the result, 2.

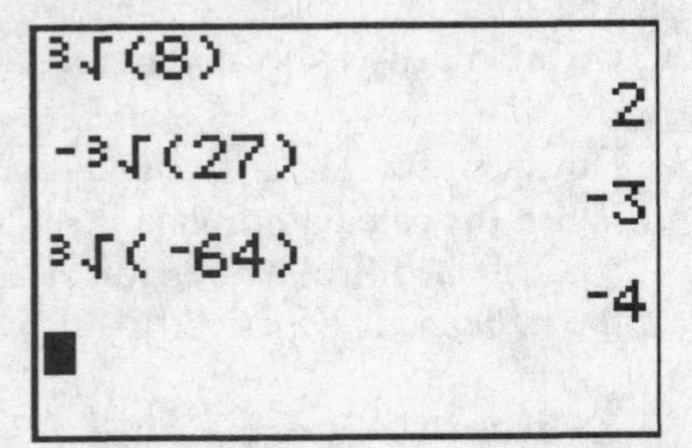

- To evaluate $-\sqrt[3]{27}$, enter the negative first, then the cube root symbol. Enter the value, 27. Press [ENTER] to see the result, -3.
- To evaluate $\sqrt[3]{-64}$, enter the cube root symbol first. Enter the value, -64. Press [ENTER] to see the result, -4.

Evaluate Roots Higher than 3

The radicands of roots higher than 3 can be evaluated using the $\sqrt[x]{\ }$ function in the MATH menu.

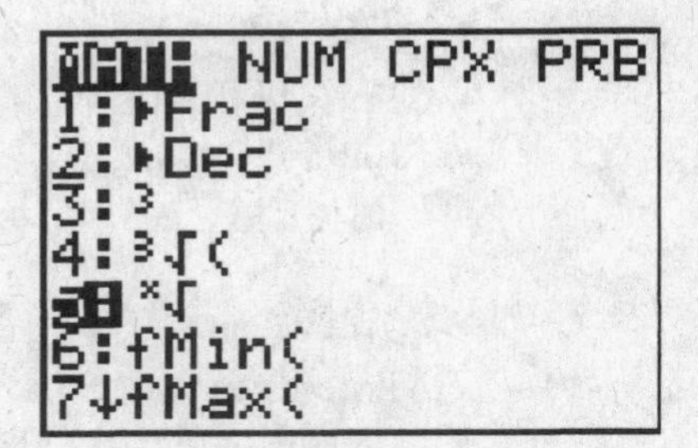

Evaluate $\sqrt[4]{16}$, $-\sqrt[5]{32}$, $\sqrt[5]{-243}$.

- To evaluate $\sqrt[4]{16}$, first enter the root you want, 4, then press the [MATH] key.
- Move the cursor to 5: $\sqrt[x]{\ }$ and press [ENTER], or press [5] to enter the root symbol on the Home screen.
- Enter the value, 16. Press [ENTER] to see the result, 2.

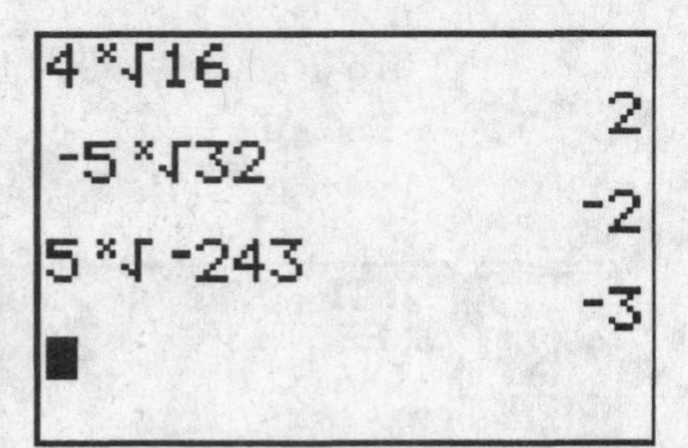

- To evaluate $-\sqrt[5]{32}$, enter the negative first, then the 5 for the fifth root.
- Press the [MATH] key. Move the cursor to 5: $\sqrt[x]{\ }$ and press [ENTER], or press [5] to enter the root symbol on the Home screen.
- Enter the value, 32. Press [ENTER] to see the result, -2.
- To evaluate $\sqrt[5]{-243}$, enter the 5, then the root symbol. Enter the value, -243. Press [ENTER] to see the result, -3.

Radicals and Rational Exponents

To evaluate with rational exponents, you may use the root notation and the definition of fractional powers, or use parentheses around a rational exponent.

Evaluate $(25)^{\frac{3}{2}}$ using root notation.

- $(25)^{\frac{3}{2}}$ can be written as $(\sqrt{25})^3$. On the Home screen enter $\sqrt{\ }(25)^\wedge 3$. Press [ENTER] to see the result, 125.

Evaluate $(25)^{\frac{3}{2}}$ using power notation.

- On the Home screen enter $(25)^\wedge(3/2)$. Press [ENTER] to see the result, 125.

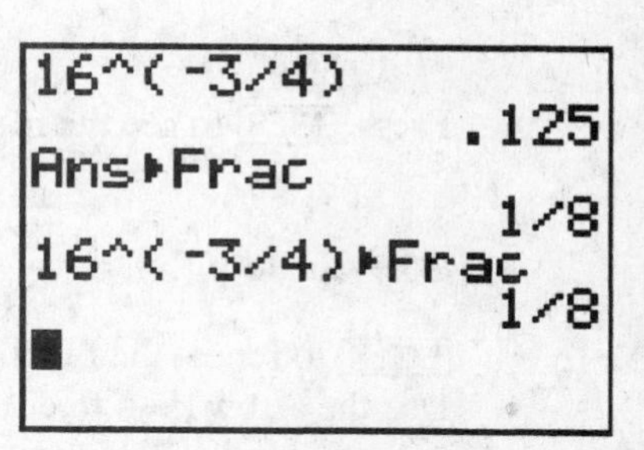

Evaluate $(16)^{-\frac{3}{4}}$ using power notation. Convert the result to a fraction if possible.

- On the Home screen enter $(16)^\wedge(-3/4)$. Press [ENTER] to see the result, .125.
- To convert the result to a fraction, press the [MATH] key and select 1:FRAC to see the result, $1/8$.
- Note that this can all be done on one line as shown.

16^(-3/4)
.125
Ans▶Frac
1/8
16^(-3/4)▶Frac
1/8

Evaluate a Radical with more than one term in the Radicand

When a radical function has more than one term under the radical, you must use parentheses at the beginning and end of the expression that is under the radical.

For $f(x)=\sqrt{3x-1}$, find $f(3)$. Round the result to two decimal places.

- Set the MODE to FLOAT 2. This setting will display all results with two decimal places.
- Press [MODE].
- Move the cursor to the line with the word FLOAT.
- Move the cursor over the number 2.
- Press [ENTER].
- To evaluate $f(x)=\sqrt{3x-1}$ when $x=3$, enter $\sqrt{\ }(3*3-1)$. Press [ENTER] to see the result, 2.83.

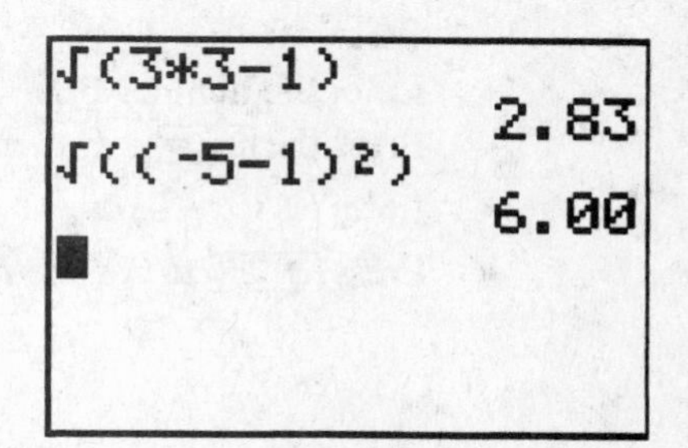

For $g(x)=\sqrt{(x-1)^2}$, find $g(-5)$. Round to two decimal places.

- To evaluate $g(x)=\sqrt{(x-1)^2}$ when $x=-5$, enter $\sqrt{\ }((-5-1)^2)$.
- Press [ENTER] to see the result, 6.00.
- Please note that you must have two sets of parentheses here to get the correct result.

For $h(x)=\sqrt[3]{8x-8}$, find $h(2)$, $h(0)$. Round to two decimal places.

```
³√(8*2-8)
                    2.00
³√(8*0-8)
                   -2.00
```

- To evaluate $h(x)=\sqrt[3]{8x-8}$ when $x=2$, enter $\sqrt[3]{\ }(8*2-8)$.
- Press [ENTER] to see the result, 2.00.
- To evaluate $h(x)=\sqrt[3]{8x-8}$ when $x=0$, enter $\sqrt[3]{\ }(8*0-8)$.
- Press [ENTER] to see the result, -2.00.

For $f(x)=\sqrt{\frac{3x}{2}}$, find $f(72), f(120)$. Round to one decimal place.

- Set the MODE to FLOAT 1. This setting will display all results with one decimal place.
- To evaluate $f(x)=\sqrt{\frac{3x}{2}}$ when $x=72$, enter $\sqrt{\ }(3*72/2)$.
- Press [ENTER] to see the result, 10.4.

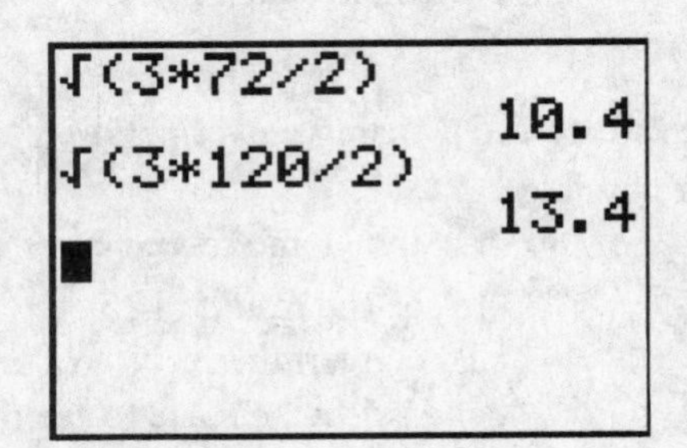

- To evaluate $f(x)=\sqrt{\frac{3x}{2}}$ when $x=120$, press [2nd] [ENTER] to repeat the function line.
- Use the arrow keys to change the 72 to 120.
- Note that you must enter an extra digit.
- Use INS (INSERT) above the [DEL] key to do this.
- Press [ENTER] to see the result, 13.4.

Use [STO▸] to evaluate a function at several values

For $f(x)=\sqrt[3]{x-3}$, find $f(30)$, $f(11)$, $f(-122)$.

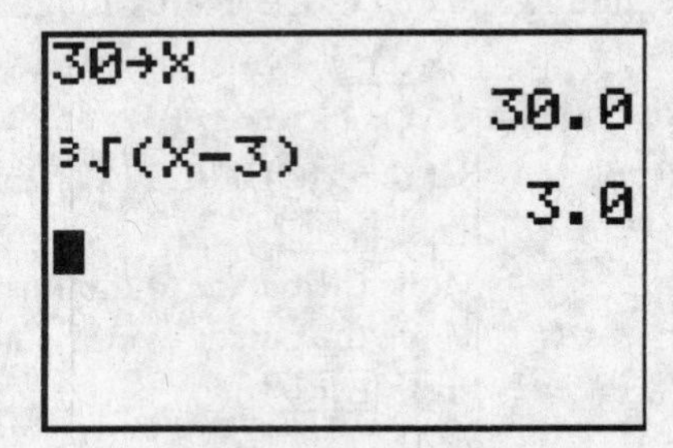

- To evaluate $f(x)=\sqrt[3]{x-3}$ when $x=30$, first store 30 to x.
- Enter the number 30.
- Press the [STO▸] key then [X,T,Θ,n] [ENTER].
- Enter $\sqrt[3]{\ }(x-3)$.
- Press [ENTER] to see the result, 3.0.

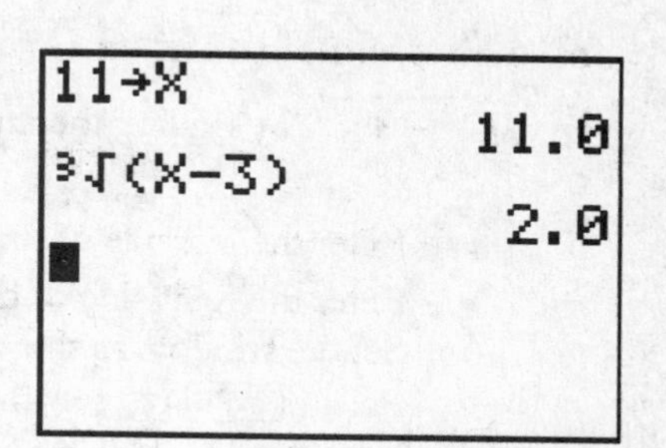

- To evaluate $f(x)=\sqrt[3]{x-3}$ when $x=11$, first store 11 to x.
- Enter the number 11.
- Press the [STO▸] key then [X,T,Θ,n] [ENTER].
- Press [2nd] [ENTER] (ENTRY) until you find the expression $\sqrt[3]{\ }(x-3)$ again.
- Press [ENTER] to see the result, 2.0 .

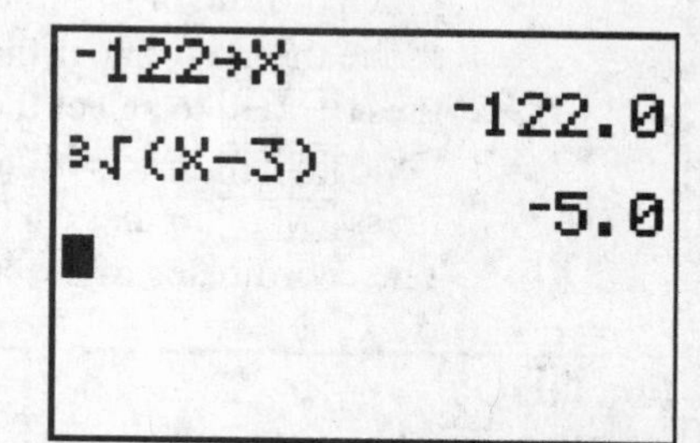

- To evaluate $f(x)=\sqrt[3]{x-3}$ when $x=-122$, first store -122 to x.
- Press the [STO▸] key then [X,T,Θ,n] [ENTER].
- Press [2nd] [ENTER] (ENTRY) until you find the expression $\sqrt[3]{\ }(x-3)$ again.
- Press [ENTER] to see the result, -5.0 .

Use a Table to evaluate a function at several values

Evaluate $f(x)=2.9\sqrt{x}+20.1$ at $x=\{0,6,12,18,24,36,48,60\}$.

- Set the MODE to CONNECTED.
- Enter the function in the Y= menu.
- Press the [Y=] key.
- Enter the function as Y1= $2.9\sqrt{\ }(x)+20.1$.
- Press [2nd] [WINDOW] (TBL SET) to set the table to Indpnt: ASK.
- Use the arrow keys to move the cursor over the word ASK.
- Press [ENTER].
- Press [2nd] [GRAPH] (TABLE).
- Start entering the values of x in the first column.
- Note that to see the last value at $x=60$, you will be writing over the last line and lose its values.
- Record the values found.

X	Y1
0.0	20.1
6.0	27.2
12.0	30.1
18.0	32.4
24.0	34.3
36.0	37.5
48.0	40.2

X=48

X	Y1
0.0	20.1
6.0	27.2
12.0	30.1
18.0	32.4
24.0	34.3
36.0	37.5
60.0	42.6

X=60

Solve Radical Equations Graphically using Intersection

Solve $\sqrt{2x-3}=9$ by finding the intersection of the left and right sides of the equation.

- Enter the left side of the equation as Y1=$\sqrt{2x-3}$.
- Enter the right side of the equation as Y2= 9 .
- Set the window as shown below.
- Press [GRAPH] to view the graphs.
- Use [2nd] [TRACE] (CALC) to identify the intersection point.
- Move the cursor to 5:intersect and press [ENTER], or press [5]
- Move the cursor near the intersection point.
- Press [ENTER] to select the first curve.
- Press [ENTER] to select the second curve.
- Press [ENTER] to make a GUESS.
- The coordinates of the intersection point are displayed at the bottom of the screen.

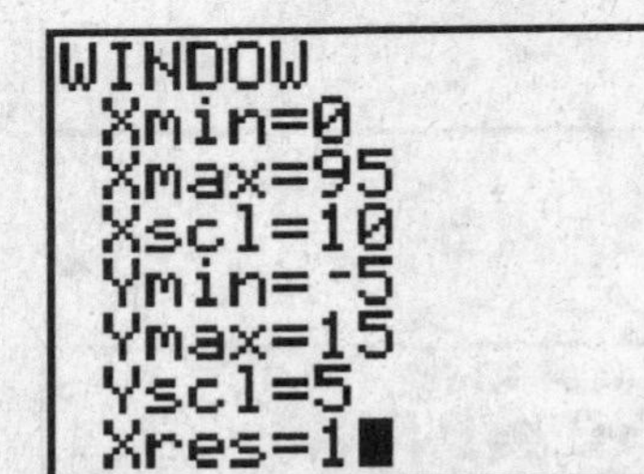

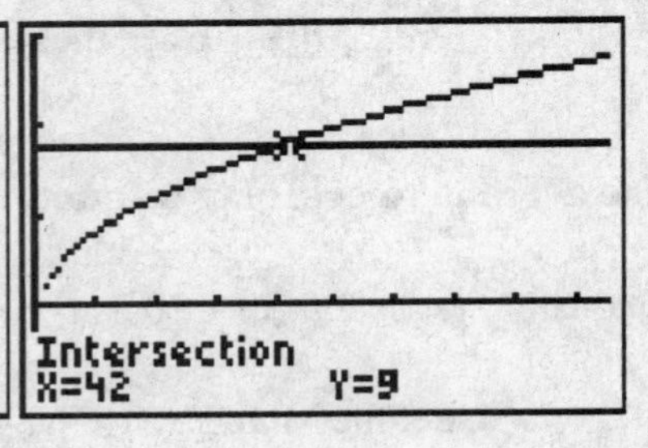

Solve Radical Equations Graphically using the x-intercept

Solve $\sqrt{2x-3}=9$ by finding the x-intercepts of $\sqrt{2x-3}-9=0$.

- Enter the left side of the equation as Y1= $\sqrt{\ }(2x-3)-9$.
- Set the window as shown in the previous example above.
- Press [GRAPH] to view the graphs.
- Use [2nd] [TRACE] (CALC) to identify the zero. Select 2:zero. Press [ENTER].
- Move the cursor slightly to the left of the x-intercept and press [ENTER].
- Move the cursor slightly to the right of the x-intercept and press [ENTER].
- Press [ENTER] again to find the coordinates of the x-intercept or Zero displayed at the bottom of the screen.

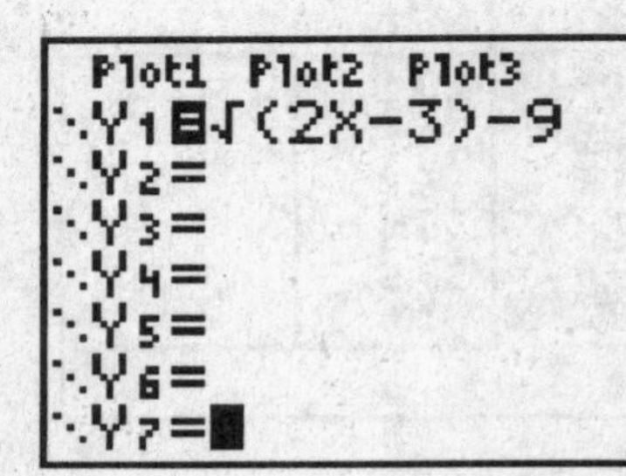

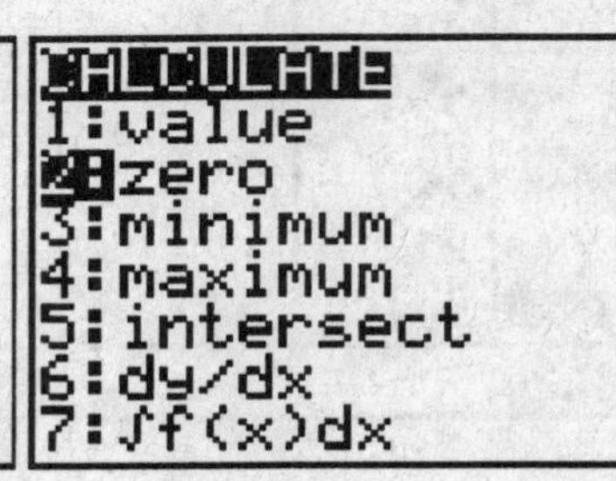

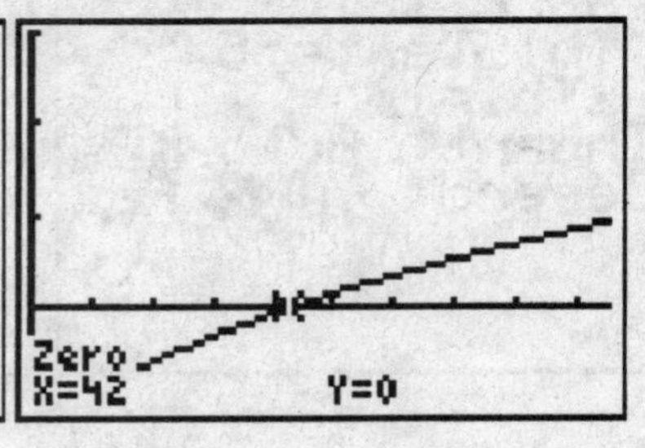

COMPLEX NUMBERS $a+bi$

Complex number arithmetic may be performed on both the TI-83 PLUS and TI-84 PLUS calculators. Set the MODE to $a+bi$ as shown.

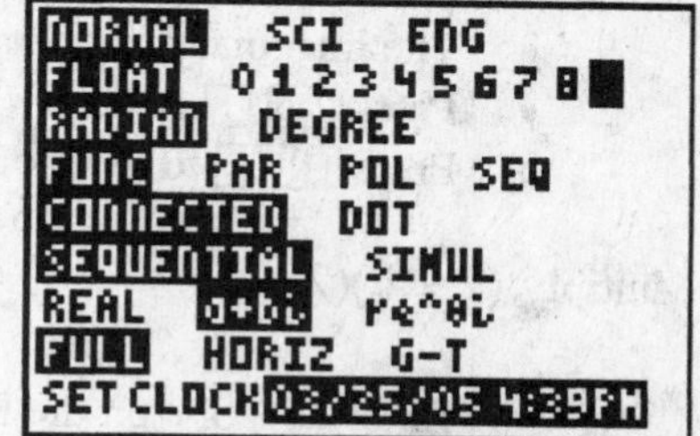

- Press the [MODE] key.
- Move the cursor to the line that begins with the word REAL
- Move the cursor to the right, and place it over $a+bi$.
- Press [ENTER].
- This setting will display complex numbers in $a+bi$ form.

Evaluate Complex Numbers on the Home Screen

Evaluate $\sqrt{-9}$, $-\sqrt{-25}$, $\sqrt{-2}\cdot\sqrt{-8}$.

```
√(-9)
                3i
-√(-25)
               -5i
√(-2)*√(-8)
                -4
```

- Set the MODE to $a+bi$ as shown above.
- Enter the values on the Home screen given as shown.

Add $(4-7i)+(2+3i)$

- Enter $(4-7i)+(2+3i)$ on the Home screen.
- The imaginary part i can be found as the second function over the decimal key.
- Press [2nd] [.] to enter i.
- Press [ENTER] to see the result, $6-4i$.

Subtract $(6+5i)-(8-i)$

- Enter $(6+5i)-(8-i)$ on the Home screen.
- The imaginary part i can be found as the second function over the decimal key.
- Press [2nd] [.] to enter i.
- Press [ENTER] to see the result, $-2+6i$.

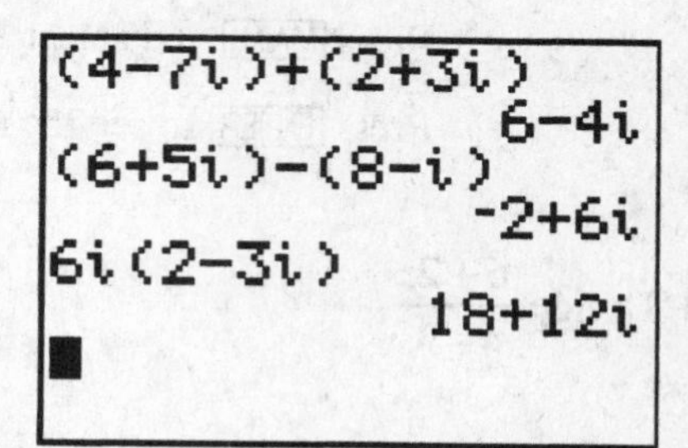

Multiply $6i(2-3i)$

- Enter $6i(2-3i)$ on the Home screen.
- The imaginary part i can be found as the second function over the decimal key.
- Press [2nd] [.] to enter i.
- Press [ENTER] to see the result, $18+12i$.

Multiply $(\sqrt{5}-5i)(\sqrt{5}+5i)$

- Enter $(\sqrt{\ }(5)-5i)(\sqrt{\ }(5)+5i)$ on the Home screen.
- The imaginary part i can be found as the second function over the decimal key.
- Press [2nd] [.] to enter i.
- Press [ENTER] to see the result, 30.

Multiply $(2-4i)(2-i)$

- Enter $(6-3i)^2$ on the Home screen.
- Press [ENTER] to see the result, $-10i$.

```
(√(5)-5i)(√(5)+5
i)
                30
(2-4i)(2-i)
              -10i
(6-3i)²
            27-36i
█
```

Multiply $(6-3i)^2$

- Enter $(6-3i)^2$ on the Home screen.
- Press [ENTER] to see the result, $27-36i$.

Divide $\dfrac{4}{i}$

- Enter $\dfrac{4}{i}$ on the Home screen.
- Press [ENTER] to see the result, $-4i$.

Divide $\dfrac{9}{1-2i}$

- Enter $\dfrac{9}{1-2i}$ on the Home screen.
- Press [MATH] and select 1: ➔Frac to see the result in fraction form.
- Press [ENTER] to see the result, $\dfrac{9}{5}+\dfrac{18}{5}i$.

```
4/i
               -4i
9/(1-2i)▸Frac
        9/5+18/5i
(6+2i)/(4-3i)▸Fr
ac
     18/25+26/25i
█
```

Divide $\dfrac{6+2i}{4-3i}$

- Enter $\dfrac{6+2i}{4-3i}$ on the Home screen.
- Press [MATH] and select 1: ➔Frac to see the result in fraction form.
- Press [ENTER] to see the result, $\dfrac{18}{25}+\dfrac{26}{25}i$.

CHAPTER 14

QUADRATIC EQUATIONS AND FUNCTIONS

SOLVE QUADRATIC EQUATIONS GRAPHICALLY

Solve Quadratic Equations Graphically using Intersection

Solve $(x+2)^2 = 18$.

```
Plot1 Plot2 Plot3
\Y1=(X+2)²
\Y2=18
\Y3=
\Y4=
\Y5=
\Y6=
\Y7=
```

- Enter the left side of the equation as Y1= $(x+2)^2$
- Enter the right side as Y2= 18.
- Set the window as shown below.
- Press [GRAPH] to view the graphs.
- Press [2nd] [TRACE] (CALC).
- Select 5:intersect.
- Move the cursor near the intersection point.
- Press [ENTER] to select the first curve.
- Press [ENTER] to select the second curve.
- Press [ENTER] to select the make a guess and see the coordinates of the intersection point at the bottom of the screen.
- The first intersection point is at $(-6.242641, 18)$.
- Repeat the process to find the second intersection point at $(2.2426407, 18)$.

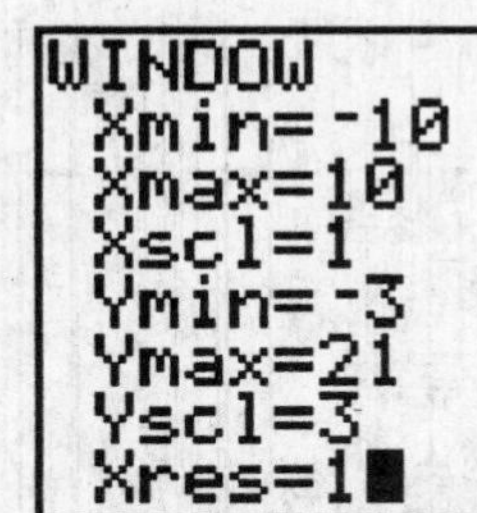

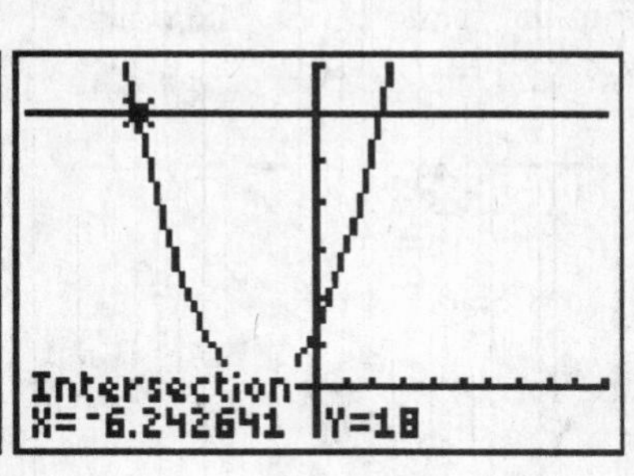

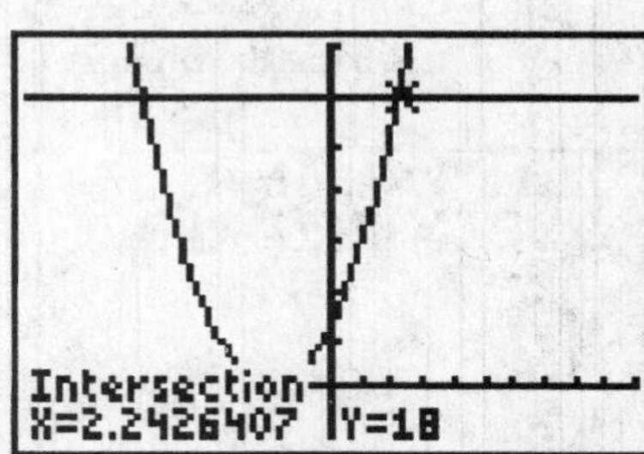

Check the solution on the Home Screen

When this quadratic equation, $(x+2)^2 = 18$, is solved algebraically using the square root property, the solution is found to be $-2 \pm 3\sqrt{2}$. Verify that the algebraic solution is the same as the solutions found using Intersection.

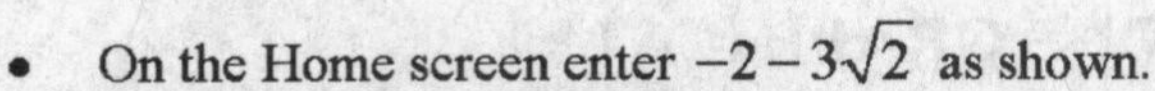

- On the Home screen enter $-2-3\sqrt{2}$ as shown.
- Press [ENTER] to see the result, -6.242640687.
- Press [2nd] [ENTER] (ENTRY) to repeat the line.
- Change the connecting sign to +.
- Press [ENTER] to see the result, 2.242640687.
- Note that the results are the same as the intersection points found by graphing in the first example.

Solve Quadratic Equations Graphically using x-intercepts

Solve $(x+2)^2 = 18$ by solving $(x+2)^2 - 18 = 0$.

- Enter the left side of the equation as Y1= $(x+2)^2 - 18$.
- Set the window as shown below.
- Press [GRAPH] to view the graphs.
- Use [2nd] [TRACE] (CALC) to identify the zero.
- Select 2:zero. Press [ENTER].
- Move the cursor slightly to the left of the x-intercept and press [ENTER].
- Move the cursor slightly to the right of the x-intercept and press [ENTER].
- Press [ENTER] again to find the x-intercept at the point $(-6.242641, 0)$.
- Repeat the process to find the second x-intercept at the point $(2.2426407, 0)$

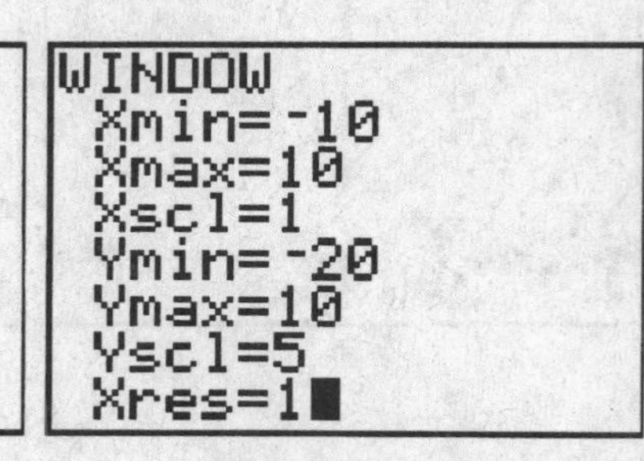

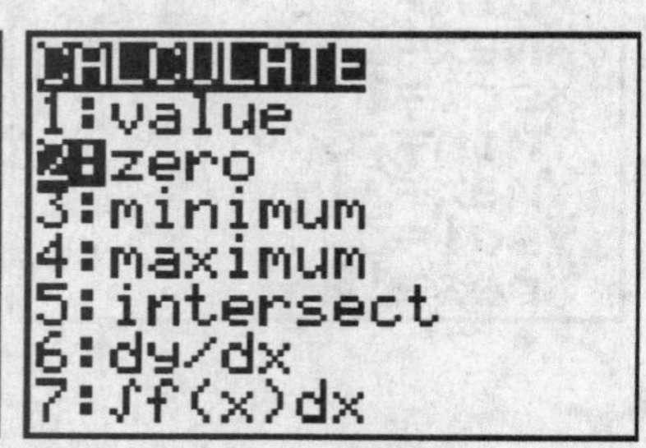

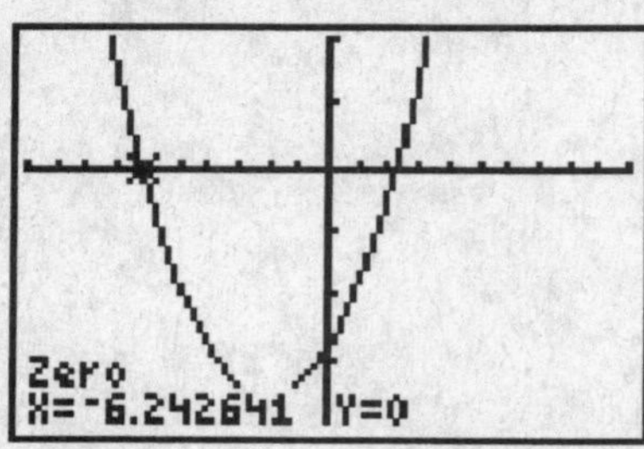

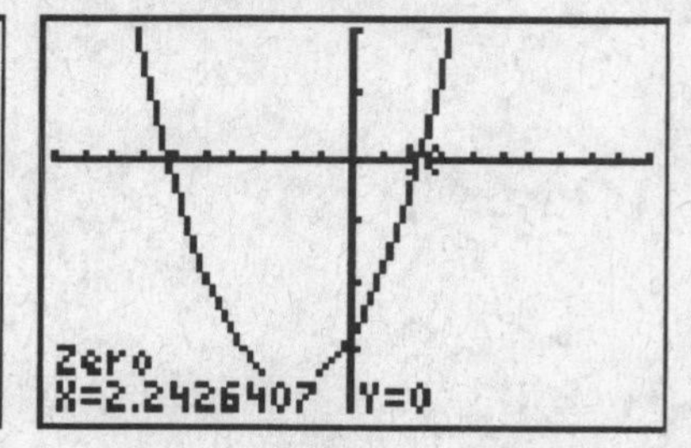

Check the solution on the Home Screen by using the Y= name of the function

- Recall that the equation $(x+2)^2-18=0$ was entered as Y1= $(x+2)^2-18$.
- Y1= $(x+2)^2-18$ can be evaluated on the Home Screen by using the VARS key.
- Press VARS.
- Use the right arrow to move to Y-VARS.
- Select 1:Function.
- Select 1:Y1. This puts Y1 on the Home screen.
- Use function notation $\text{Y1}(-6.242641)$ to evaluate the Y1 at the first solution, $x=-6.242641$.
- Press ENTER to see the result, $2.654881\,E-6$ or 0.000002654881, which is effectively 0.
- Repeat for the second solution.
- Enter $\text{Y1}(2.2426407)$ to see the result $1.09296E-7$ or 0.00000010926, which is effectively 0.
- Recall that the exact solutions are $-2\pm3\sqrt{2}$.
- Repeat the evaluation of Y1 using the exact values of $-2\pm3\sqrt{2}$ as shown.
- Note that the result is exactly 0.

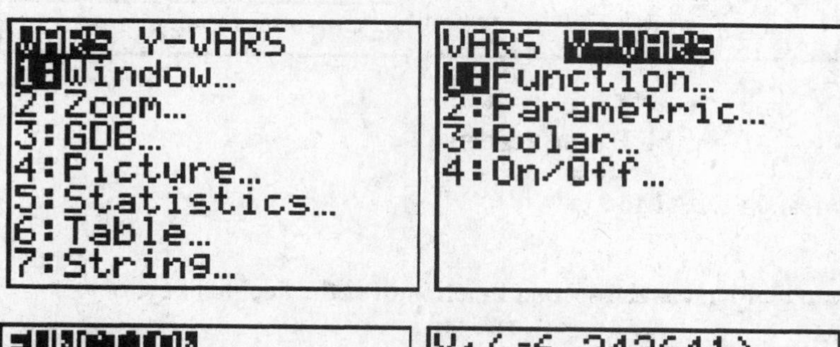

Complex Solutions of a quadratic equation

Complex number arithmetic may be performed on both the TI-83 PLUS and TI-84 PLUS calculators. Set the MODE to $a+bi$ as shown below.

Graph $y=3x^2+6x+5$. Check the complex number solutions on the Home screen.

- Enter the function as $\text{Y1}=3x^2+6x+5$.
- Use the window shown to graph the function.
- Note that the entire graph is above the x-axis, so there is no real value at which $y=0$.
- Note that zero feature that you find in the CALC menu only finds real zeros, even if you have set the mode to $a+bi$.

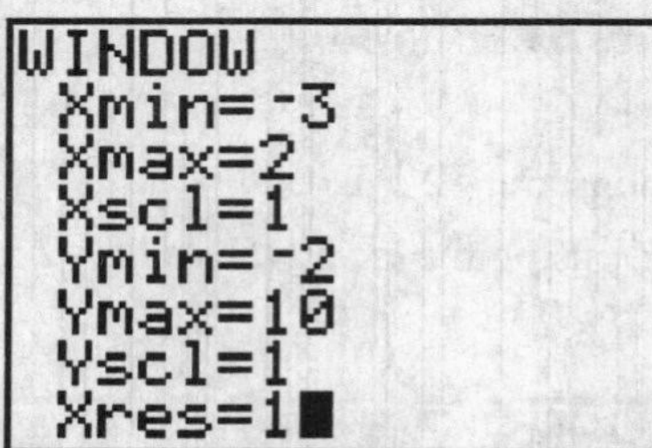

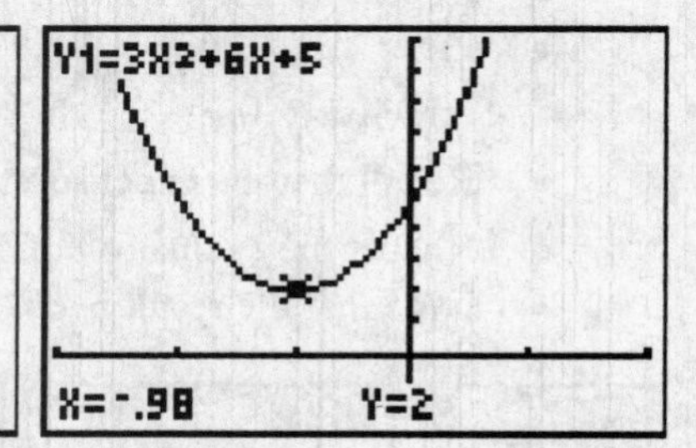

Check the algebraic results $-1\pm\frac{\sqrt{6}}{3}i$ on the Home screen

- Set the MODE to $a+bi$ and FLOAT to 2.
- Enter $-1+\sqrt{(6)}/3i$.
- The imaginary part i can be found as the second function over the decimal key.
- Press [2nd] [.] to enter i.
- Press the [STO▸] [X,T,Θ,n] to store $-1+\sqrt{(6)}/3i$ in the variable x.
- Note that the parentheses around the radicand are necessary.
- Note that you can do this on one line as shown.
- Next, enter the expression $3x^2+6x+5$ and press [ENTER] to see the result, 0.00.
- Repeat the process for the second solution, $-1-\sqrt{(6)}/3i$.

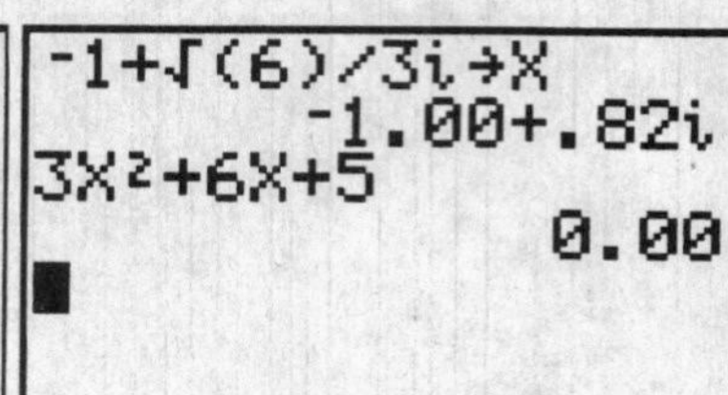

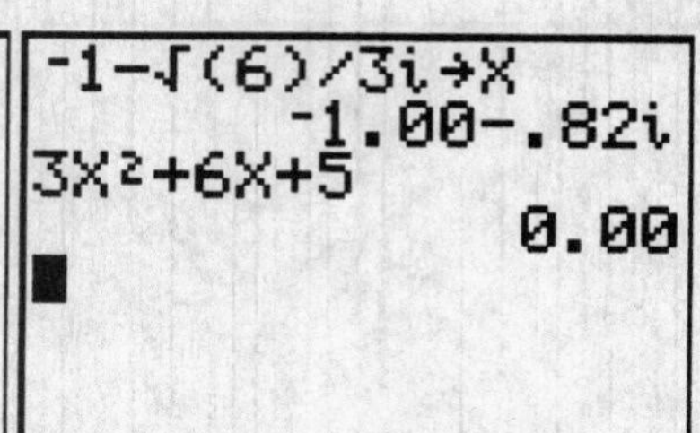

Find the vertex of a quadratic function

The vertex of a quadratic function is also called the maximum or minimum point of the function. The TI-83 Plus and the TI-84 Plus give you the exact coordinates of the vertex, once you identify the vertex as a maximum or minimum of the function.

Find the vertex of the quadratic function discussed above, $y = 3x^2 + 6x + 5$.

- Enter the function as Y1=$3x^2 + 6x + 5$.
- Use the window shown below to graph the function.
- First notice from the graph that the vertex is the lowest point on the curve, so the vertex is a minimum point of the function.
- From the graph screen, press [2nd] [TRACE] (CALC).
- Select 3:minimum.
- You are directed to identify a left and right bound of the minimum as well as a Guess.
- Move the cursor somewhere to the left of the vertex and press [ENTER].
- Move the cursor somewhere to the right of the vertex and press [ENTER].
- You can enter a value for the guess or just press [ENTER].
- The x and y coordinates of the minimum point, or vertex of this function, are shown at the bottom of the screen.
- The vertex of the function $y = 3x^2 + 6x + 5$ is the point $(-1, 2)$.

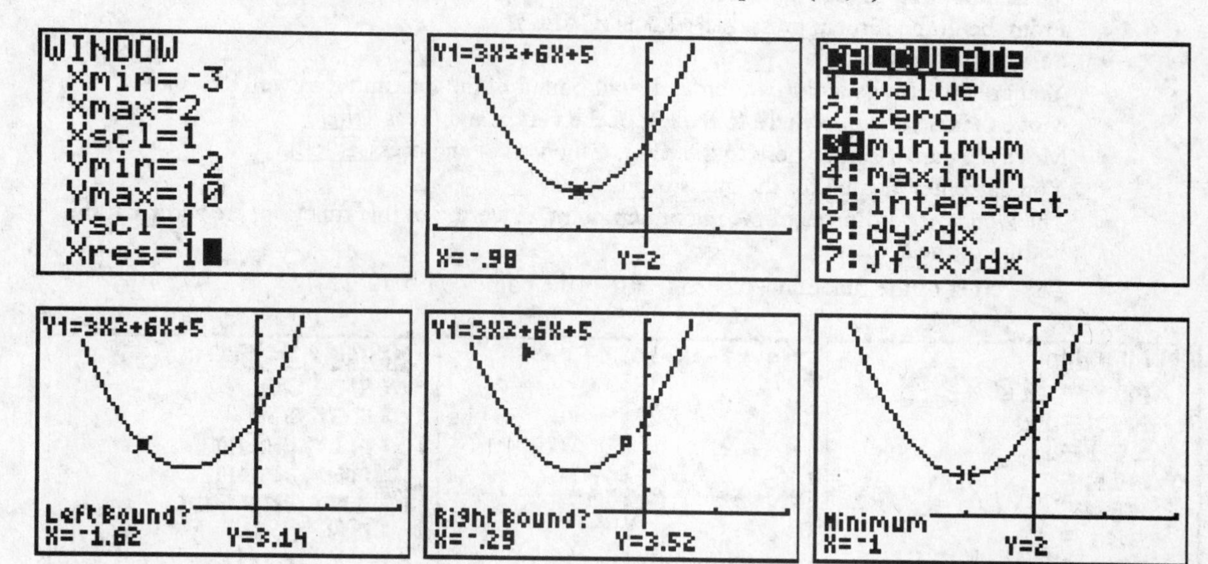

Find the x-intercepts of $y = 3x^2 + 6x + 5$.

- Note that the graph does not touch the x-axis, so there are no real x-intercepts. Therefore, there are no real solutions to the equation $3x^2 + 6x + 5 = 0$.
- There are, however, two complex number solutions to the equation $3x^2 + 6x + 5 = 0$.

Find the y-intercept of $y = 3x^2 + 6x + 5$.

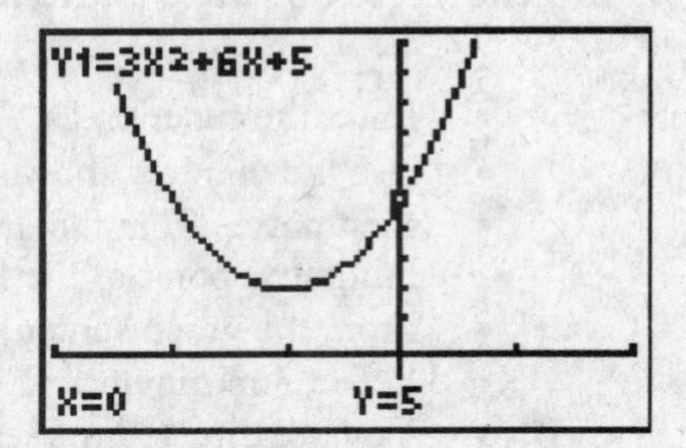

- The y-intercept is the point on the graph that crosses the y-axis, when $x = 0$.
- From the graph screen, press [TRACE].
- Enter the number 0 to evaluate Y1 when $x = 0$.
- Press [ENTER] to see the y-coordinate of this function is $y = 5$ when $x = 0$.

Find the vertex of the quadratic function $y = -x^2 - 3x + 10$.

- Enter the function as $\text{Y1} = -x^2 - 3x + 10$.
- First notice from the graph that the vertex is the highest point on the curve, so the vertex is a maximum point of the function.
- From the graph screen, press [2nd] [TRACE] (CALC).
- Select 4:maximum.
- You are directed to identify a left and right bound of the maximum as well as a Guess.
- Move the cursor somewhere to the left of the vertex and press [ENTER].
- Move the cursor somewhere to the right of the vertex and press [ENTER].
- You can enter a value for the guess or just press [ENTER].
- The x and y coordinates of the maximum point, or vertex of this function, are shown at the bottom of the screen.
- The vertex of the function $-x^2 - 3x + 10$ is the point $(-1.5, 12.25)$.

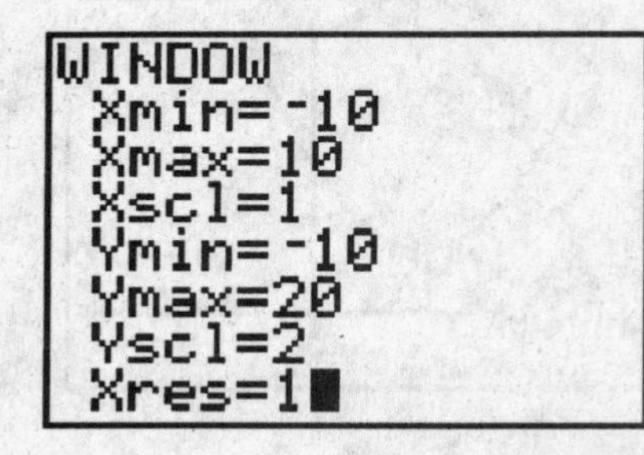

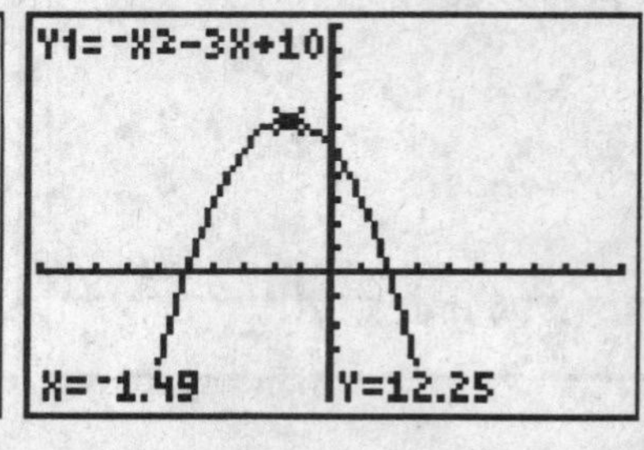

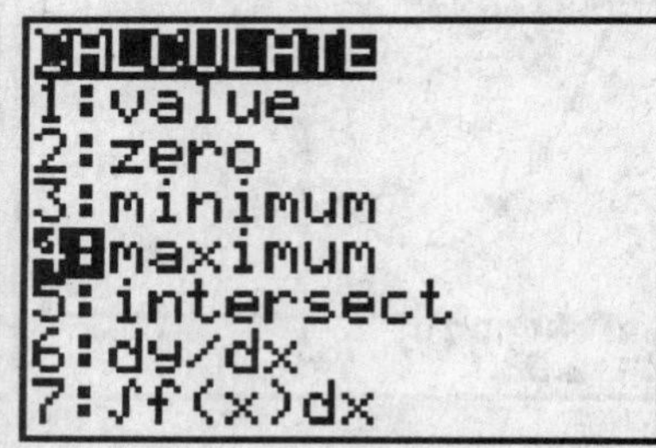

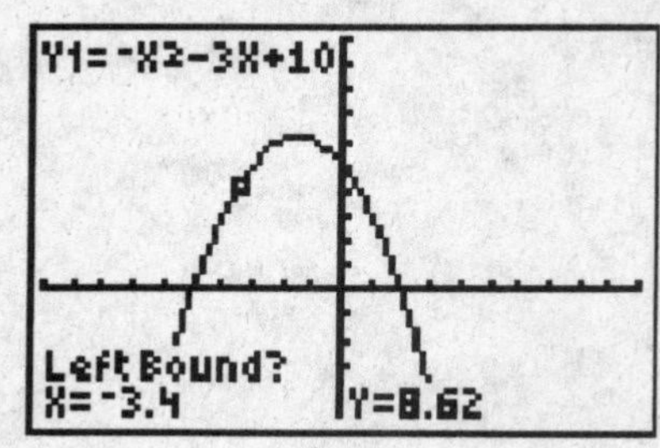

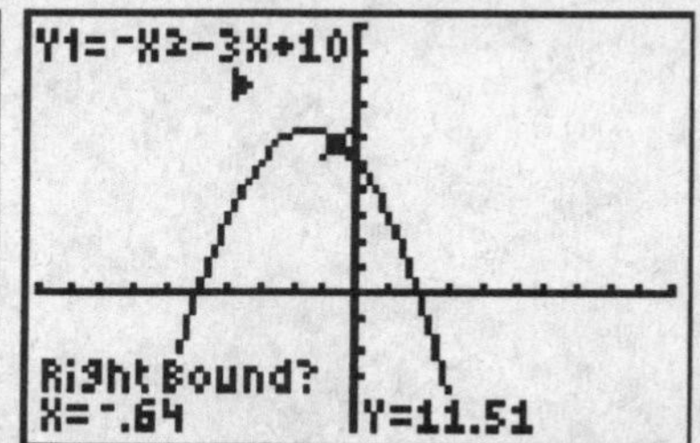

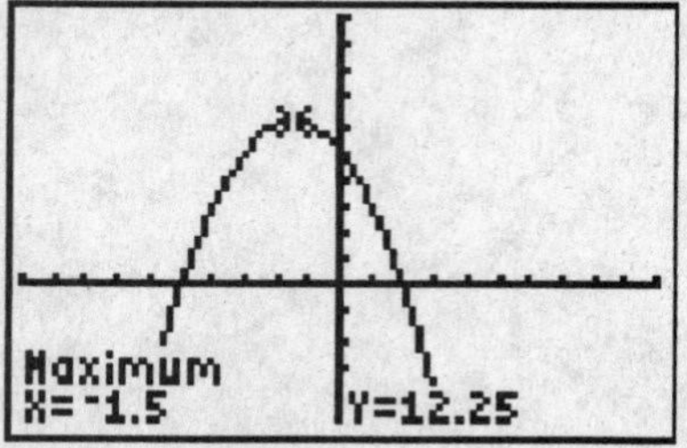

Find the x-intercepts of $y = -x^2 - 3x + 10$.

- Note that the graph has two x-intercepts.
- To find the first x-intercept, press [2nd] [TRACE] (CALC).
- Select 2:zero.
- Move the cursor slightly to the left of the first x-intercept.
- Press [ENTER].
- Move the cursor slightly to the right of the first x-intercept.
- Press [ENTER].
- Press [ENTER] guess.
- The x and y coordinates of the first x-intercept are displayed at the bottom of the screen.
- The first x-intercept is the point $(-5, 0)$.

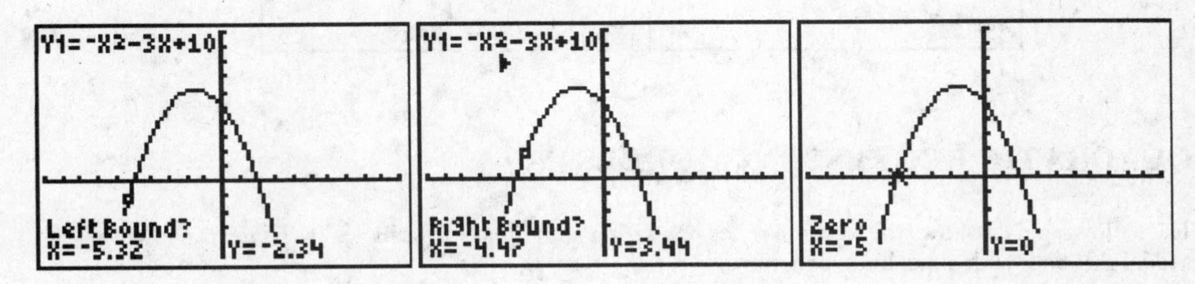

- Repeat the process to find the second x-intercept of $-x^2 - 3x + 10$ at the point $(2, 0)$.
- Press [2nd] [TRACE] (CALC).
- Select 2:zero.
- Select the left and right bounds and guess to identify the second x-intercept at the point $(2, 0)$.

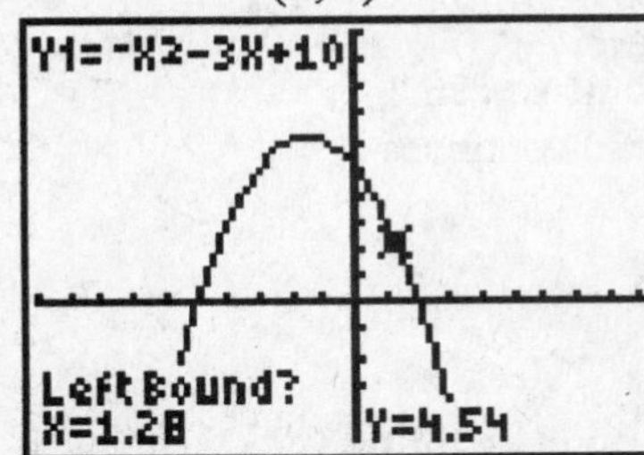

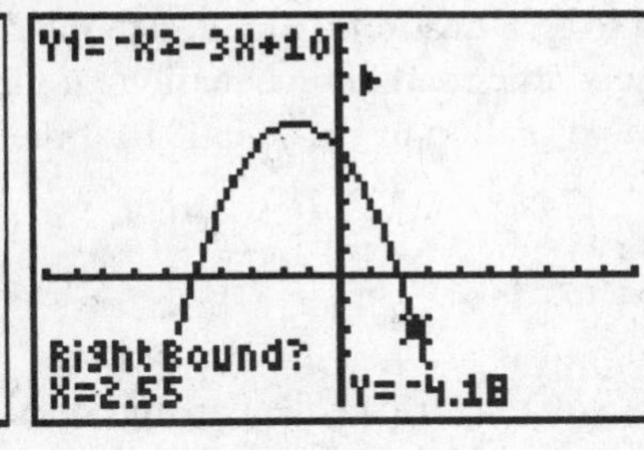

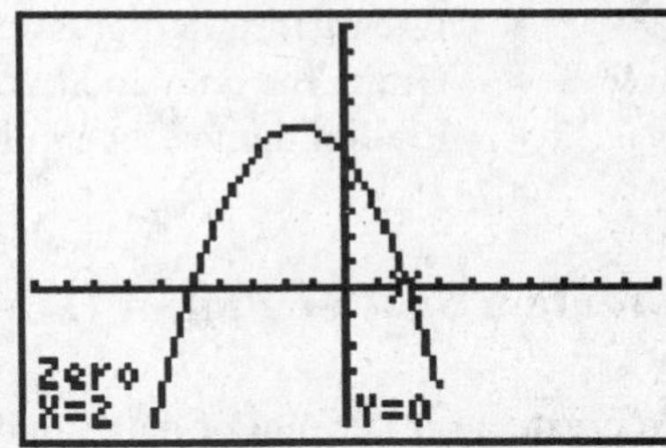

Find the y-intercept of $y=-x^2-3x+10$.

- The y-intercept is the point on the graph that crosses the y-axis, when $x=0$.
- From the graph screen, press [TRACE], then the number [0] key.
- Press [ENTER] to see the y-coordinate of this function when $x=0$.
- The y-intercept is the point $(0,10)$.

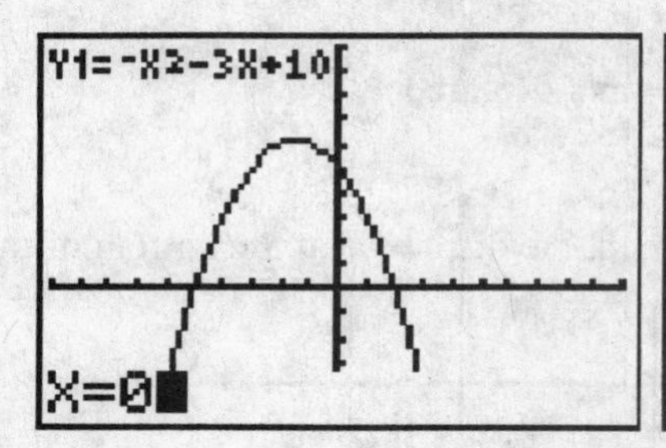

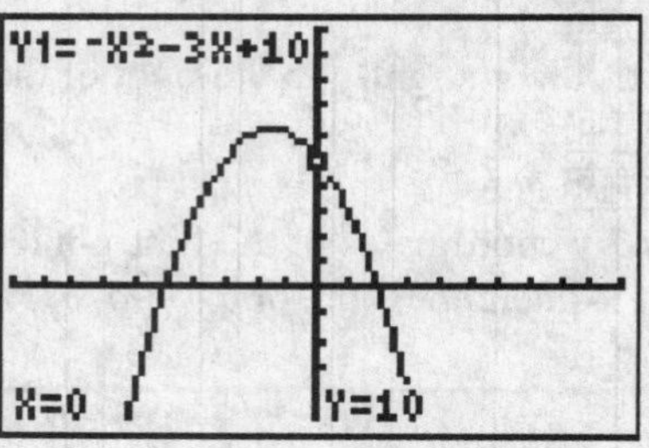

QUADRATIC REGRESSION MODELS

The following data shows the average miles per gallon MPG for U.S. vehicles in the years shown. Model this data with a quadratic equation.

YEAR	1960	1970	1980	1990	1998	2000	2003
x	0	10	20	30	38	40	43
Average Miles per gallon	13.4	12	13.2	16.2	17.1	18.5	19.8

- Plot the data points.
- Use quadratic regression to fit a quadratic function to the data.
- Graph the data and the quadratic regression equation on the same screen.
- Predict the average miles per gallon in the year 2010 if the trend continues.

Create a Scatter Plot of Data Points

To create a scatter plot of data points, you must first select an appropriate window for the data. Then you must enter the data as lists, use the STAT menu to define Plot 1, Plot 2, or Plot 3 and finally graph the data points.

Set the Window

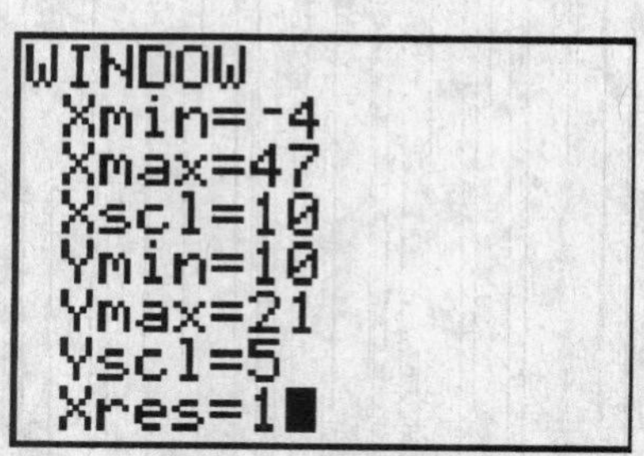

- Identify the lowest and highest values of x and y, then select a number about 5 or 10 more on each end. In this example, x goes 0 to 43, and y goes from 13.2 to 19.8. I would suggest a window on x as [-4, 47, 10] and on y as [10, 21,5].
- Press the [WINDOW] key. Set the window as shown.

Clear Lists

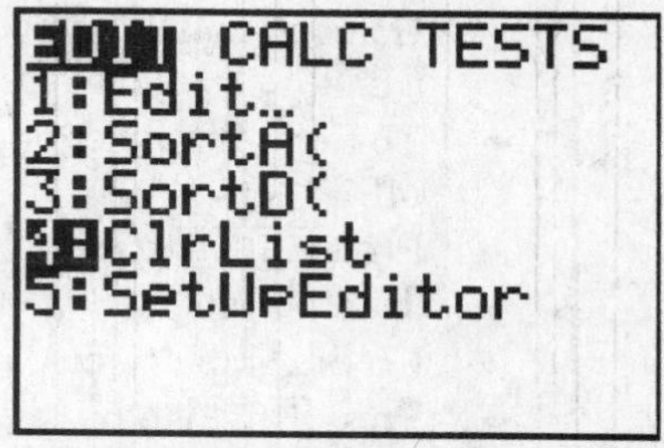

- If you wish to clear the lists, go back to STAT EDIT, and select 4: ClrList. This pastes ClrList on the Home Screen.
- Press [2nd] [1] (L_1) [ENTER] to clear L_1. You should see the word Done.
- Repeat the line by pressing [2nd] [ENTER] (ENTRY), and move the cursor over L_1.
- Press [2nd] [2] (L_2) to clear L_2.

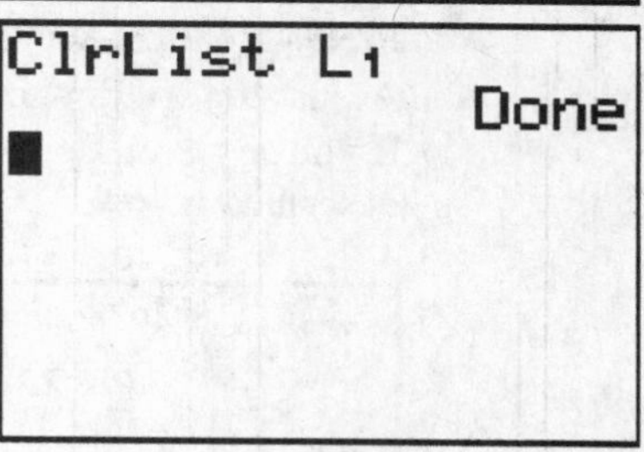

- Another way to clear entries from a list is to overwrite the elements of the list or delete any elements you want using the [DEL] key.
- Note that the [CLEAR] key does not erase a highlighted entry.

- An entire list can be cleared in the STAT menu by pressing the up arrow until the top of the list is highlighted.
- Press [CLEAR] [ENTER] in this position to clear the contents of the entire list.
- If you press the [DEL] key when the column heading is highlighted, it will remove the list from the STAT editor, but the list will still possess its contents. The list name can be returned to the editor with [2nd] [DEL] (INS) command and typing in the list name.

Enter the Data Points

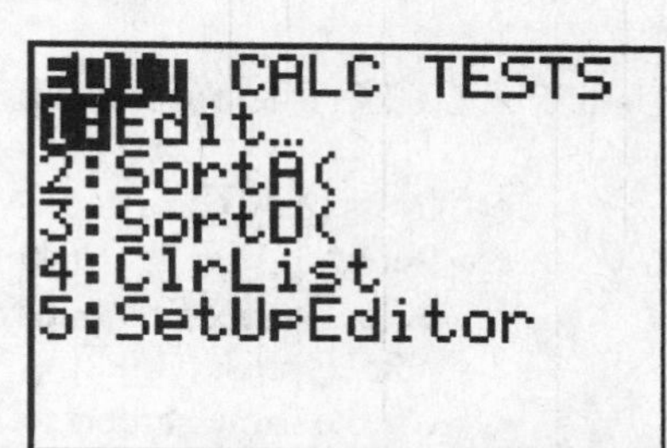

- Press the [STAT] key.
- Press [1] to select 1:Edit or move the cursor to 1:Edit and press [ENTER].
- This brings you to the screen where you can enter the data in the lists.
- You may need to clear the lists.
- Enter the Years from 1960-values in list L_1.
- Enter the MPG-values in L_2.
- Use the down arrow to move from line to line.

L1	L2	L3 1
0	13.4	------
10	12	
20	13.2	
30	16.2	
38	17.1	
40	18.5	
43	19.8	

L1(1)=0

Turn on a Plot

- Go back to the Y= menu and turn off or clear all the Y= functions, namely the ones that have a dark rectangle over the equal sign.
- To turn off a function, move the cursor over the *equal* sign of the function you wish to turn off.
- Press [ENTER] to remove the highlighted (dark) rectangle over the equal sign.
- Turn on Plot 1.
- On the graph screen, use the up arrow to go to Plot 1.
- Press [ENTER] to highlight it.
- Press [GRAPH] to see the points.
- Adjust the window if you wish to see the points more clearly.
- If you see something else, like a histogram, you must format the Plot using STAT PLOT as described below.

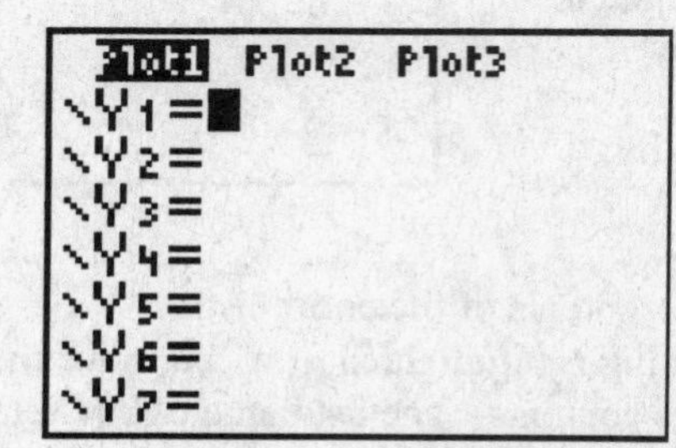

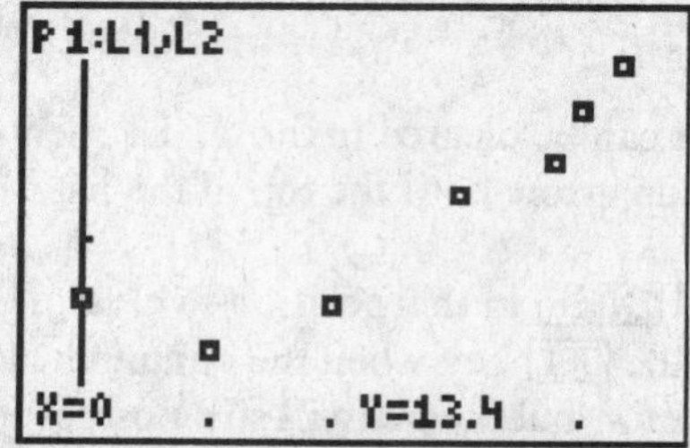

Use STAT PLOT to format the graph of data points

- Press [2nd] [Y=] to access STAT PLOT.
- Select 1: Plot 1…ON,press [ENTER].
- Select Type: first one for a scatter plot – individual points, press [ENTER].
- On the same screen select
  XList: L_1 or whichever list has the *x*-data
  YList: L_2 or whichever list has the *y*-data
  Mark: any one you wish
- Press the [GRAPH] key to see the scatter plot.
- Use [TRACE] and the right arrow to move from point to point. The *x* and *y* coordinates of each point are displayed at the bottom of the screen.
- P1:L1,L2 is displayed in the upper left hand corner of the screen. This tells you that you are on Plot 1, the *x*'s are from list L_1, and the *y*'s are from list L_2.

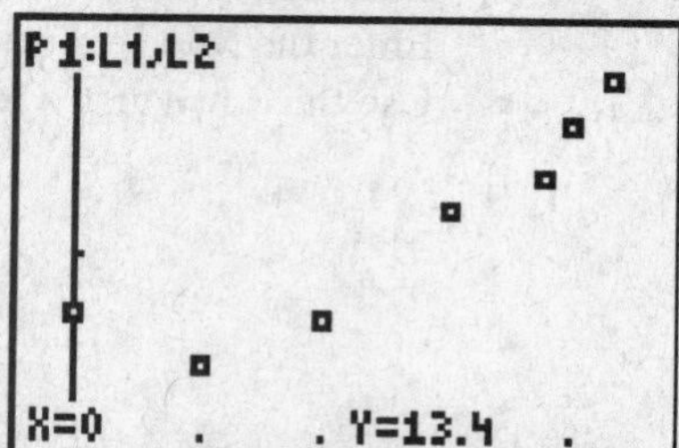

Find and graph the quadratic Regression Equation that is the "Best Fit" for the Data.

- The data is already entered into L1 and L2, and Plot 1 is formatted for scatter plot.
- Turn off or clear all Y= functions.
- Press the [STAT] key.
- Use the right arrow to go to CALC.
- Press [5] to select 5:QuadReg. This puts QuadReg on the Home screen.
- Press [ENTER] to see the values of a and b as shown.
- If you do not see r and r^2, see the directions below on how to turn Diagnostics On.
- Hence, the quadratic regression equation through the data points is $y = .006234834x^2 - .1147312317x + 13.1043665$.
- Use the [MODE] key and Float 3 to round these values to 3 decimals as shown.
- Press [CLEAR] to return to the Home screen.
- Press [ENTER] to see the Quadratic Regression equation in 3 decimal places, $y = .006x^2 - .115x + 13.104$.

Display r and r^2

You may not see the values for the linear correlation coefficient r. To do so, you must go to the Catalog and set DiagnosticOn.

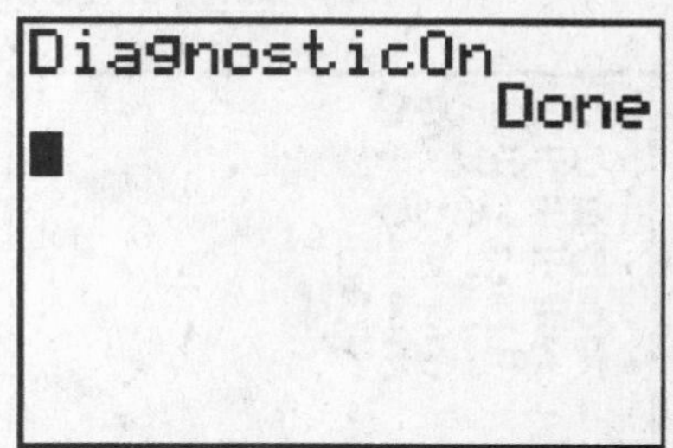

- Press [2nd] [0] to access the catalog. This is a complete listing of all the functions available on the calculator.
- You should be in alpha mode (see the A in the upper right hand corner of the screen.
- If you do not see an A in the upper right corner of the screen, press [2nd] [ALPHA] to lock Alpha.
- Press the [x^{-1}] to jump to the D's.
- Use the down arrow to find DiagnosticOn. Be sure that the arrow is next to it.
- Press [ENTER] twice until you see the word Done on the Home Screen.
- Redo finding the quadratic equation. The screen will now also include r and r^2.

Graph the Quadratic Regression Equation and Scatter Plot from the Home Screen

- Turn off all Y= functions. Be sure that Plot 1 is on.
- Press [2nd] [MODE] (QUIT) to go to the Home screen.
- Press the [STAT] key and use the right arrow to go to CALC.
- Press [5] to select 5:QuadReg.
- This puts QuadReg on the Home screen.
- Enter the names of the lists to use for the quadratic regression equation.
- Press [2nd] [1] (L1) comma [,] [2nd] [2] (L2) comma [,] .

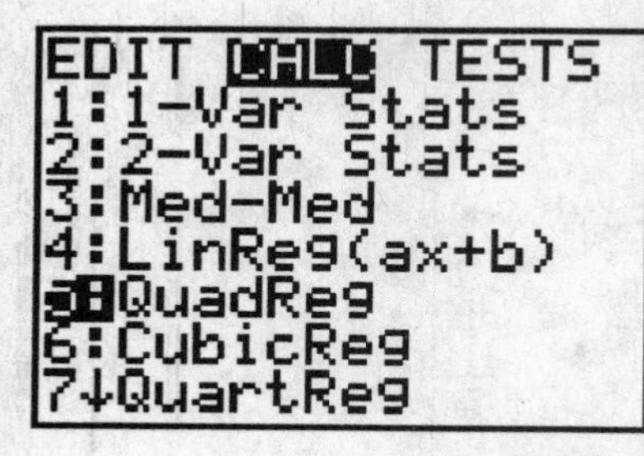

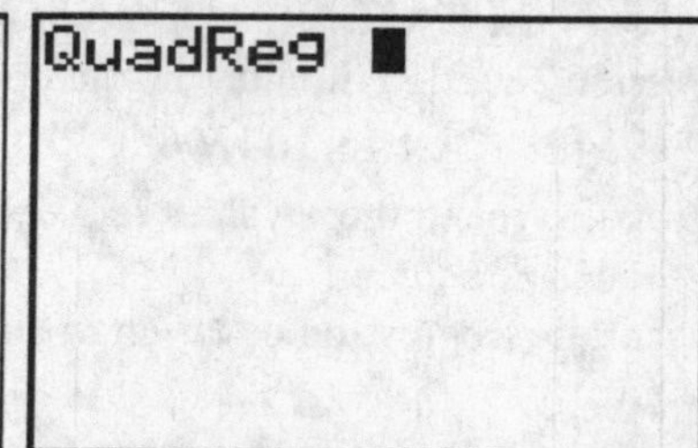

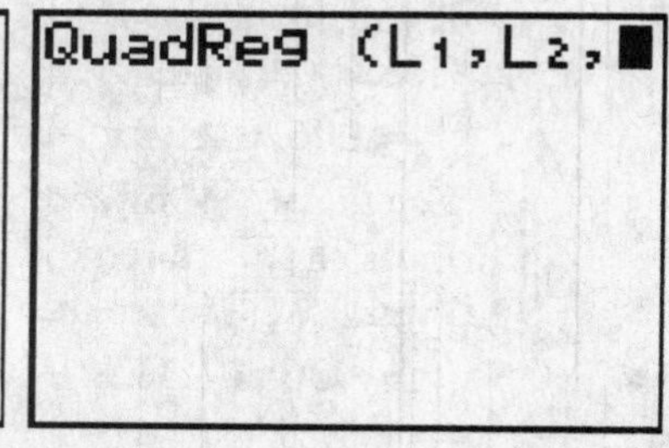

- Now you must find Y1 in the VARS menu.
- Press the [VARS] key.
- Use the right arrow to go to Y-VARS.
- Select 1:Function.
- On the Function screen, select the position in the Y=menu you wish to use for the regression equation, say Y1.

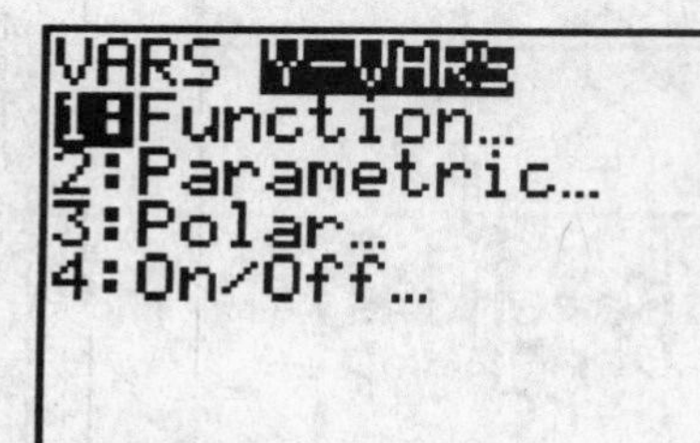

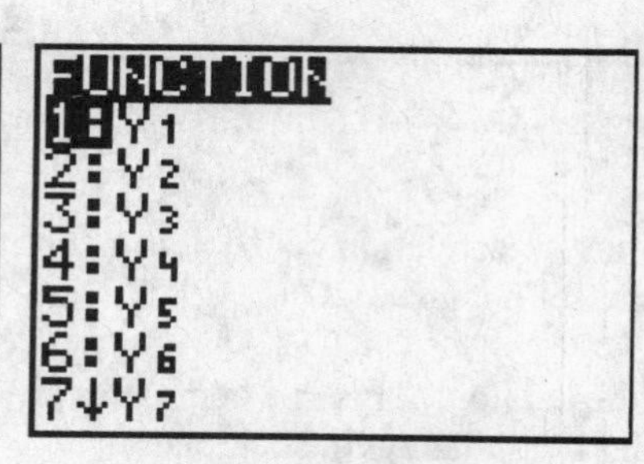

- Press [ENTER] to see the quadratic regression information on the Home screen.
- Press [Y=] to see it in the Y= menu.
- Press [GRAPH] to see the scatter plot and the quadratic regression equation on the same screen.

QuadReg
y=ax²+bx+c
a=.006
b=-.115
c=13.104
R²=.967

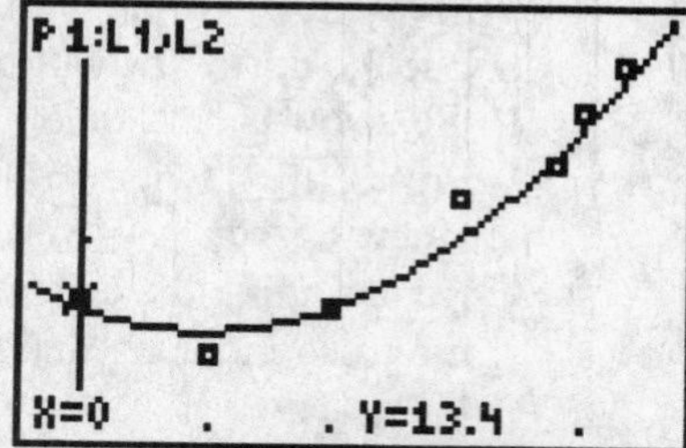

Graph the Quadratic Regression Equation and Scatter Plot from the Graph Screen

- Turn off all Y= functions.
- Be sure that Plot 1 is on.
- Move the cursor to the Y= line you wish to use.
- Press the [VARS] key.
- Select 5:Statistics.
- Move the right arrow twice to see the EQ menu.
- Select 1:RegEQ.
- Press [ENTER].
- This puts the regression equation into the Y= menu.
- Press the [GRAPH] key to see the regression equation and the scatter plot on the same screen.
- Press [TRACE] to trace the graphs.
- Use the up or down arrow to toggle between the scatter plot and the regression equation.

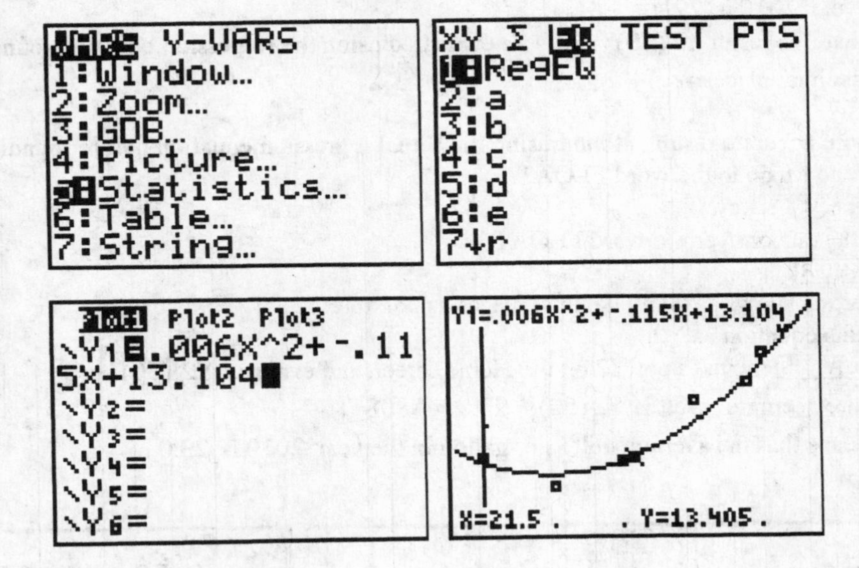

Use the Quadratic Regression Equation to predict future average Miles Per Gallon

- The year 2010 is 50 years after 1960.
- You must find the value of the linear regression equation when $x = 50$.
- Recall that the regression equation is stored as Y1.
- Use the [VARS] menu to put Y1 on the Home screen.
- Press the [VARS] key.
- Use the right arrow to go to the Y-VARS menu.
- Select 1:Function
- Select 1:Y1.
- This puts Y1 on the Home screen.
- Enter (50) to see Y1(50).
- Press [ENTER] to see the predicted value, 22.4, which means that the average miles per gallon in the year 2010 is 22.4.
- Note that this result Y1(50) = 22.4 is evaluated using the regression equation rounded to three decimal places.

Y1(50)
22.4

- The more accurate result is found using the actual regression equation before rounding.
- Re-set the Mode to the word FLOAT.
- Press the [MODE] key.
- Move the cursor over the word FLOAT.
- Press [ENTER].
- This setting shows as many decimal places as possible.
- Enter this equation as Y2.
- Use the [VARS] menu to put Y2 on the Home screen and evaluate Y2 (50).
- The more accurate result is Y2(50) = 22.95488081.
- This means that the average miles per gallon in the year 2010 is 23.0.

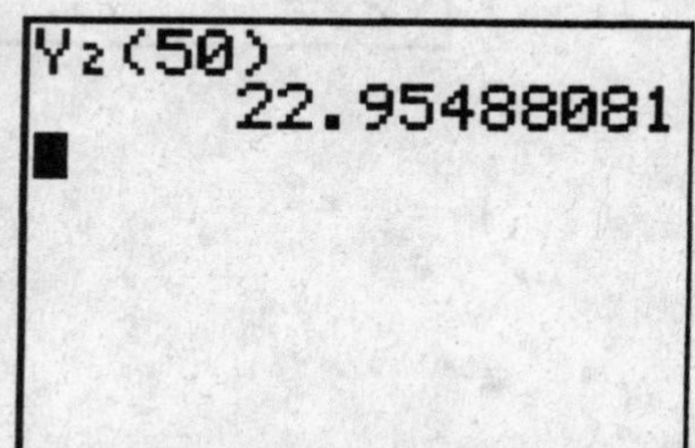

CHAPTER 15

EXPONENTIAL AND LOGARITHMIC FUNCTIONS

EXPONENTIAL FUNCTIONS

Evaluate Exponential Expressions on the Home Screen

You may evaluate an exponential expression one x-value at a time, you may use a list, or you may use a table.

Evaluate $f(x)=2^x$ for $x=25$.

- On the Home screen enter $2^\wedge 25$.
- Press [ENTER] to see the result, 33554432.

Evaluate $f(x)=2^x$ for $x=\{-3,-2,-1\}$ using a list, say L_1.

- First you must clear the list L_1. Press the [STAT] key. Select EDIT 4.ClrList.
- Press [ENTER] to put ClrList on the Home screen.
- On the same line enter the name of the list you wish to clear. To clear the list L_1, enter [2nd] Y (L_1).
- Press [ENTER] to see the word DONE. (Chapter 8 Linear Regression for more about clearing lists.)

- Next, you must enter the values in the list. Press [STAT], select EDIT 1:Edit. This brings you to the screen where you can enter the values of L_1.
- Press [2nd] [MODE] (QUIT) to return to the Home screen.
- Enter $2^\wedge$ L_1 as shown.
- Use the [MATH] menu to convert the resulting list to fractions.

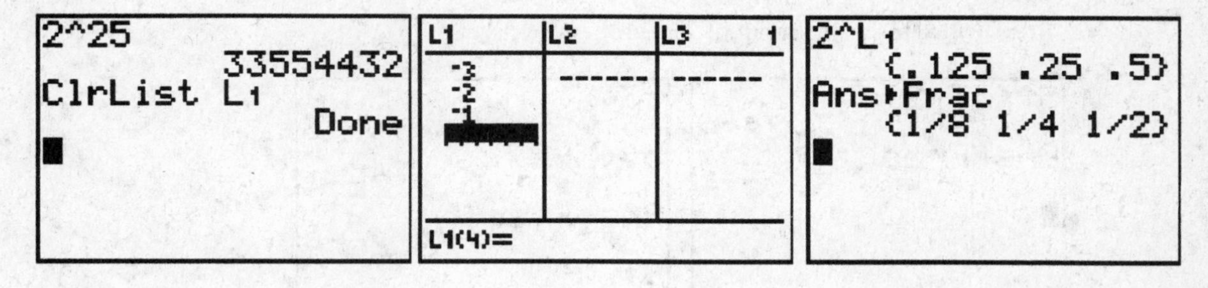

Evaluate $f(x) = 2^x$ for $x = \{-3, -2, -1, 0, 1, 2, 3\}$ using a table.

- Enter $f(x) = 2^x$ as $Y1 = 2 \wedge x$.
- Press [2nd] [WINDOW] (TBLSET) to set the table as shown to start at $x = -3$ and go by steps of 1 unit.
- Press [2nd] [GRAPH] to view the table.

Evaluate $f(x) = \left(\frac{2}{3}\right)^x$ at $x = -3$ on the Home screen.

- Enter $(2/3) \wedge -3$ as shown.
- Press [ENTER] to see the result, 3.375.
- Use the [MATH] menu to convert the result to a fraction.

Evaluate $f(x) = 5^{x-2}$ at $x = -3$ on the Home screen.

- Enter $5 \wedge (-3-2)$ as shown.
- Press [ENTER] to see the result, $3.2\text{E}-5$.
- Note that this value does not convert to a fraction.

- You could compute this using [STO▸].
- Enter -3 [STO▸] [X,T,Θ,n]. Press [ENTER].
- Enter the expression $5 \wedge (x-2)$.
- Press [ENTER] to see the result, $3.2\text{E} -4$.

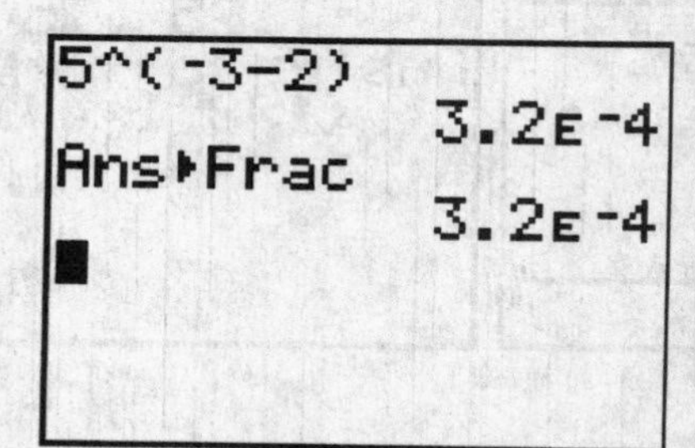

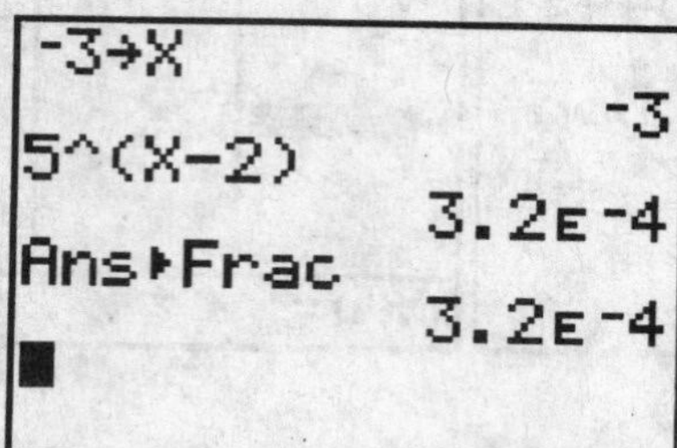

Evaluate the Compound Interest Formula $A = P\left(1+\frac{r}{n}\right)^{nt}$

Evaluate the compound interest formula for A when $P = 2000$, $r = .05$, $n = 4$, $t = 7$.

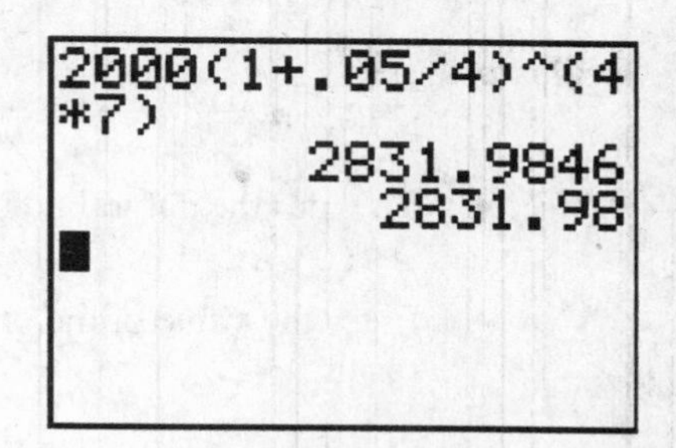

- On the Home screen enter the formula as shown.
- Note that the parentheses around the product $(4*7)$ in the exponent are essential.
- Because this is a money question, change the MODE to FLOAT 2.
- Press [ENTER] to the result in dollars and cents.

Evaluate the Compound Interest Formula using APPS

Evaluate the compound interest formula for A when $P = 2000$, $r = .05$, $n = 4$, $t = 7$.

- Press the APPS key.
- Select 1:Finance.
- Select 1:TVM Solver.
- In this application **N** is the number of periods. You can enter $4*7$ or 28.
- I % is the rate per year, which is $r = 5\%$. Enter 5.
- PV is the Present Value, or Principal P of 2000.
- You must enter 4 for compounding quarterly in P/Y, the number of payment periods per year. C/Y will automatically be the same value.
- To solve, move the cursor up to FV, the future value, A. Press the green [ALPHA] key, then [ENTER] (SOLVE) to solve the equation. Note that this is the same value that was computed on the Home screen.
- Press [2nd] [MODE] (QUIT) to return to the Home screen.

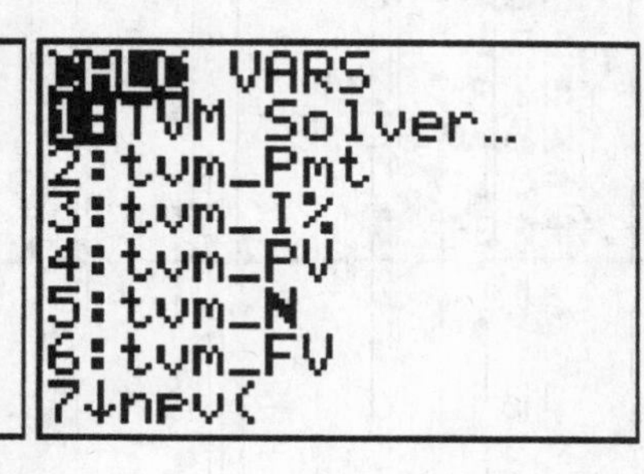

Evaluate Exponential Expressions with Base e

To evaluate with base e, you must use [2nd] [LN] e^x.

Evaluate e^2, $\frac{1}{e^3}$.

- To enter the natural number e, press [2nd] [LN] to see e^(.
- Enter the value of the exponent. Press [ENTER] to see the result.

Graph Exponential Equations

Graph $y = 800e^{0.035t}$. Find y when $t = 7$.

- Press [Y=]. Enter the equation as Y1 $= 800e$^$(0.035x)$.
- Next choose an appropriate window. Since we want to evaluate when $x = 7$, choose x as [0 , 10, 1].
- To find the range of values for y, press [TRACE], the press the [7] key and [ENTER]. This will automatically tell you the value of y when $x = 7$, namely 1022.0971. So choose the values [750, 1500, 100].

```
Plot1 Plot2 Plot3
\Y1=800e^(0.035X
)
\Y2=
\Y3=
\Y4=
\Y5=
\Y6=
```

```
WINDOW
 Xmin=0
 Xmax=10
 Xscl=1
 Ymin=750
 Ymax=1500
 Yscl=100
 Xres=1
```

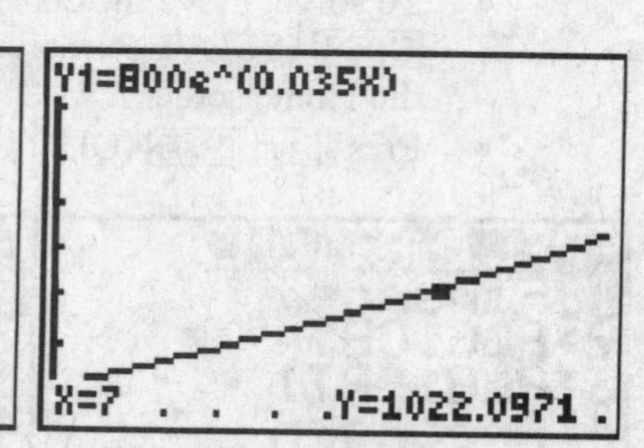

LOGARITHMIC FUNCTIONS

You may evaluate a logarithmic expression one x-value at a time, you may use a list, or you may use a table. See directions about exponential functions for using a list or a table.

Evaluate Base 10 Logarithms on the Home Screen

The logarithm whose base is 10 is called the common log and can be found on the calculator using the [LOG] key.

Evaluate $\log(10), \log(2), \log(300/3.2)$

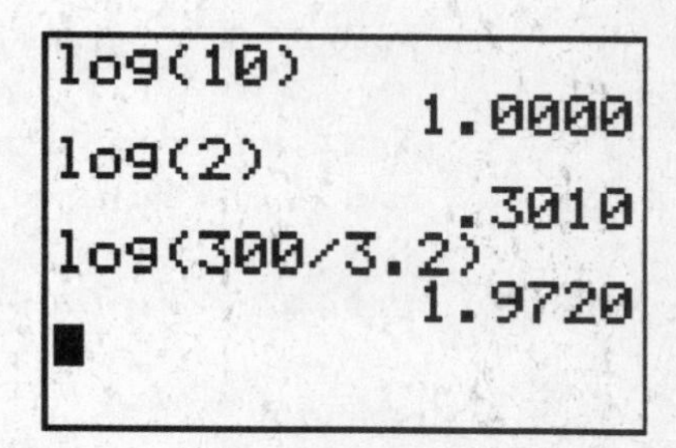

- Logarithms are often rounded to four decimal places, so set the MODE to FLOAT 4.
- For each of the values, press the [LOG] key, then enter the value as shown.
- Press [ENTER] to see the result.

Evaluate Base e Logarithms on the Home Screen

The logarithm whose base is e is called the natural log and can be found on the calculator as the [LN] key.

Evaluate $\ln(e), \ln(10), \ln(\sqrt[4]{e})$

```
ln(e^(1))
               1.0000
ln(10)
               2.3026
ln(e^(1/4))
                .2500
```

- Logarithms are often rounded to four decimal places, so set the MODE to FLOAT 4.
- For each of the values, press the [LN] key, then enter the value as shown.
- Press [ENTER] to see the result.

Evaluate $\dfrac{\log 2}{4\log 1.1025}$

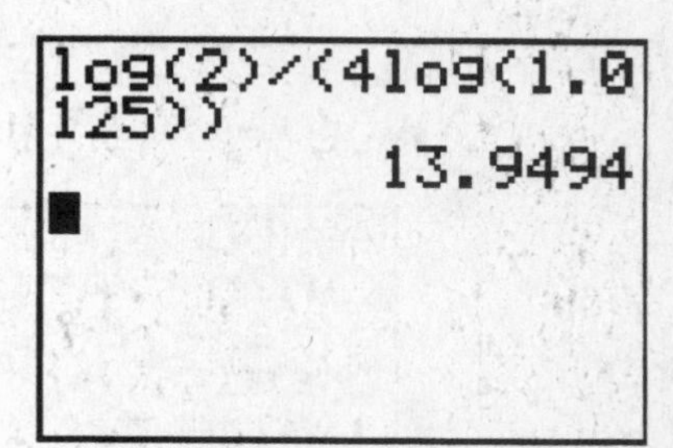

- Enter the expression as shown.
- Note that the parentheses in the denominator are essential because the denominator is a product.
- Press [ENTER] to see the result.

Using the Change of Base Formula

You may evaluate the logarithm of a base that is not 10 or e with the change of base formula and either the common or natural log. $\log_b a = \dfrac{\log a}{\log b}$ *or* $\log_b a = \dfrac{\ln a}{\ln b}$

Evaluate $\log_7 5$ using the LOG key and the LN key

- Identify that the base $b = 7$ and that the argument $a = 5$.
- Evaluate $\dfrac{\log 5}{\log 7}$ as shown.
- Note that the result is the same using the common log base 10 and the natural log base e.

Graph Logarithmic Equations

Graph $f(x) = \log(x - 2)$

- Press Y=. Enter the equation as Y1 $= \log(x - 2)$.
- The first graph on the right shows the graph in the standard window. Note that it is hard to see the behavior of the log on this screen.
- To get a better view, start with ZOOM 0:ZoomFit. Note that the Zoom Fit window has x-values between –10 and 10 and y-values between -.89 and .9.

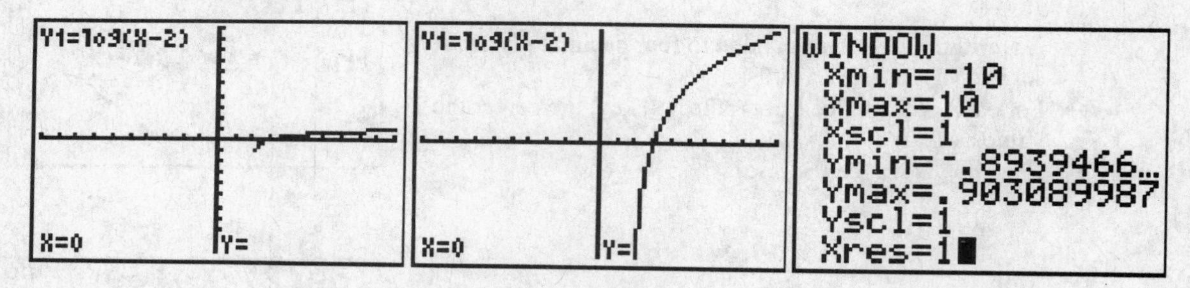

- Adjust the x and y values of the window as shown to get a better graph of the function.

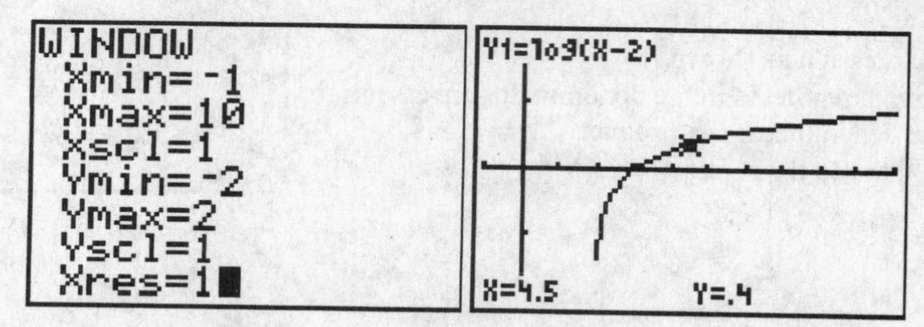

CHAPTER 16

CONIC SECTIONS

GRAPHING CONIC SECTIONS

The Parabola

Parabola of the Form $y = a(x-h)^2 + k$
The parabola in this form is a function of x. Its graph opens upward when $a > 0$. The graph opens downward when $a < 0$. Enter the function as written into Y1.

Graph $y = 5(x+5)^2 + 3$

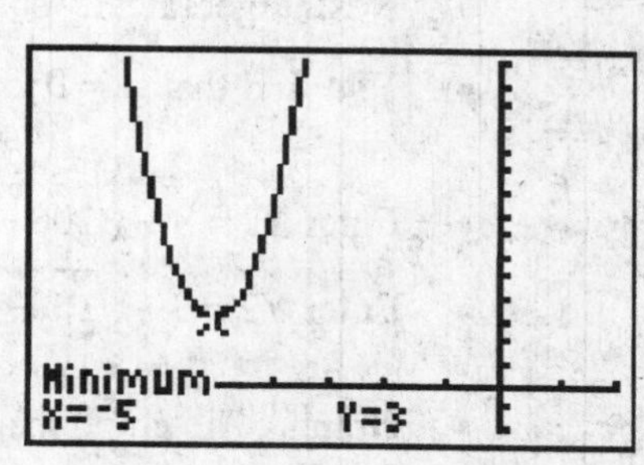

- Enter Y1 $= 5(x+5)^2 + 3$.
- Adjust the window.
- Graph.
- Use CALC 3:minimum to find the vertex.

Parabola of the Form $x = a(y-k)^2 + h$
The parabola in this form is not a function of x. You must solve the equation for y and enter the two resulting equations as Y1 and Y2. When they are both turned on, you will see the graph of the original equation. The graph of the parabola in this form opens to the right when $a > 0$ or opens to the left when $a < 0$.

Graph $x = -3(y-1)^2 + 2$

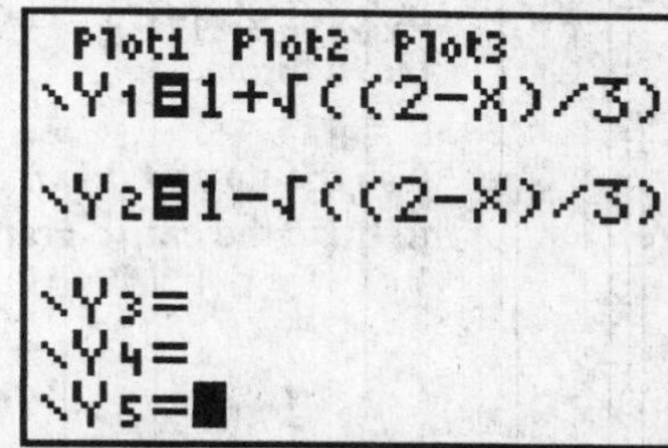

- First algebraically solve $x = -3(y-1)^2 + 2$ for y.
- You find that $y = 1 \pm \sqrt{\dfrac{x-2}{-3}}$ or $y = 1 \pm \sqrt{\dfrac{2-x}{3}}$.
- Enter $\text{Y1} = 1 + \sqrt{\dfrac{2-x}{3}}$.
- Enter $\text{Y2} = 1 - \sqrt{\dfrac{2-x}{3}}$.
- Graph.
- Use the up and down arrows to move through Y1 and Y2 to trace the entire graph.

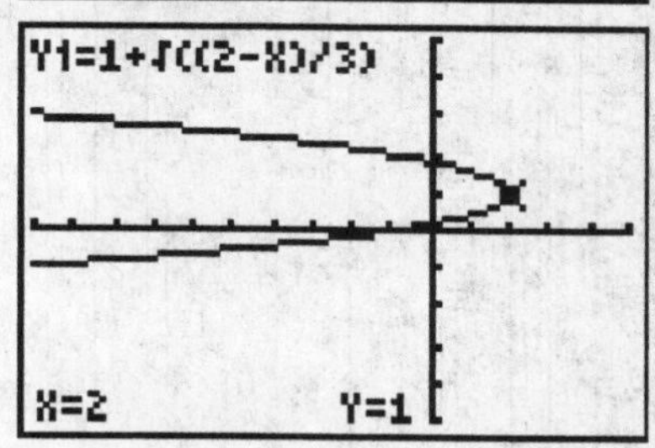

The Circle

Circle $(x-h)^2+(y-k)^2$

Circles are not functions of x. You must solve the equation for y and enter the two resulting equations as Y1 and Y2. When they are both turned on, you will see the graph of the original equation. In addition, the graph will look like a circle when you select ZOOM 5:Zsquare.

Graph $(x+7)^2+(y-3)^2=100$

- Identify the center at $(-7,3)$ and the radius $r=10$.
- Next, algebraically solve $(x+7)^2+(y-3)^2=100$ for y.
- You find that $y=3\pm\sqrt{100-(x+7)^2}$.

- Enter Y1$=3+\sqrt{100-(x+7)^2}$.
- Enter Y2$=3-\sqrt{100-(x+7)^2}$.

- Graph. The graph may not look like a circle.
- Adjust the window to see the entire graph.
- Here is a suggested window:
  [-20, 10, 5] and [-10, 20, 5].

- Press the ZOOM key.
- Select 5:Zsquare. The graph is re-drawn and looks more like a circle.

- Use the up and down arrows to move through Y1 and Y2 to trace the entire graph.

The Ellipse

Ellipse of the Form $\frac{(x-h)^2}{a^2}+\frac{(y-k)^2}{b^2}=1$

Ellipses are not functions of x. You must solve the equation for y and enter the two resulting equations as Y1 and Y2. When they are both turned on, you will see the graph of the original equation. Use ZOOM 5:Zsquare to show a graph that shows the true ellipse.

Graph $\frac{(x-1)^2}{4}+\frac{(y+2)^2}{9}=1$

- Identify the center at $(1,-2)$.
- Identify $a=3, b=2$ and that the major axis is vertical.
- Identify the vertices at $(1,-2+3)$ and $(1,-2-3)$, or at $(1,1)$ and $(1,-5)$.
- Next, algebraically solve $\frac{(x-1)^2}{4}+\frac{(y+2)^2}{9}=1$ for y.
- You find that $y=-2\pm\sqrt{\frac{36-9(x-1)^2}{4}}$.
- Enter Y1 $=-2+\sqrt{\frac{36-9(x-1)^2}{4}}$.
- Enter Y2 $=-2-\sqrt{\frac{36-9(x-1)^2}{4}}$.
- Graph. Note that graph may not look like the major axis is vertical.
- The first graph shown has window [-5, 5, 1] and [-10, 5, 1].
- Press the ZOOM key.
- Select 5:Zsquare.
- The graph is re-drawn and is the true shape of the ellipse is shown.
- Use the up and down arrows to move through Y1 and Y2 to trace the entire graph.
- As a check, evaluate Y1 and Y2 at $x=1$ to verify the vertices at $(1,1)$ and $(1,-5)$.

The Hyperbola

Hyperbola of the Form $\frac{x^2}{a^2}-\frac{y^2}{b^2}=1$

Hyperbolas are not functions of x. You must solve the equation for y and enter the two resulting equations as Y1 and Y2. When they are both turned on, you will see the graph of the original equation.

Graph $\frac{x^2}{9}-\frac{y^2}{4}=1$

- Identify the center at $(0,0)$.
- Identify that the transverse axis is horizontal.
- Identify $a=3, b=2$.
- Identify the vertices at $(-3,0)$ and $(3,0)$.
- Next, algebraically solve $\frac{x^2}{9}-\frac{y^2}{4}=1$ for y.
- You find that $y=\pm\sqrt{\frac{(4x^2-36)}{9}}$.
- Enter Y1 $=\sqrt{\frac{(4x^2-36)}{9}}$.
- Enter Y2 $=-\sqrt{\frac{(4x^2-36)}{9}}$.
- Graph. Use the up and down arrows to trace the entire graph.

Hyperbola of the Form $\frac{y^2}{a^2}-\frac{x^2}{b^2}=1$

Hyperbolas are not functions of x. You must solve the equation for y and enter the two resulting equations as Y1 and Y2. When they are both turned on, you will see the graph of the original equation.

Graph $\frac{y^2}{4}-\frac{x^2}{9}=1$

- Identify the center at $(0,0)$.
- Identify that the transverse axis is vertical.
- Identify $a=2,\ b=3$.
- Identify the vertices at $(0,-2)$ and $(0,2)$.
- First algebraically solve $\frac{y^2}{4}-\frac{x^2}{9}=1$ for y.
- You find that $y=\pm\sqrt{\frac{(4x^2+36)}{9}}$.
- Enter Y1 $=\sqrt{\frac{(4x^2+36)}{9}}$.
- Enter Y2 $=-\sqrt{\frac{(4x^2+36)}{9}}$.
- Graph. Use the up and down arrows to trace the entire graph.

CHAPTER 17

SEQUENCES

SEQUENCES

An infinite sequence is a function whose domain is the set of natural numbers $\{1, 2, 3, 4, \ldots\}$.
A finite sequence is a function whose domain is the set of natural numbers $\{1, 2, 3, 4, \ldots, n\}$.
The notation a_n denotes the nth term of the sequence.

The TI-83 PLUS and TI-84 PLUS can find a specific term of a sequence, find n terms of a sequence, graph a sequence, and display the table of a sequence.

Sequence Mode

- To put the calculator into sequence mode, press the [MODE] key.
- Move the cursor to the fourth line, which sets the graphing mode on the calculator.
- Move the cursor over the word SEQ.
- Press [ENTER].
- In the SEQ setting, the [X,T,Θ,n] will display n instead of x.

Find a Specific Term of a Sequence

Find the fourth and tenth term of the sequence, $a_n = 2^n$.

- First store the number 4 to n on the Home screen.
- Press [4] [STO▸] [X,T,Θ,n].
- Press [ENTER] to see the number 4 on the next line.
- Enter the general term of the sequence, $2 \wedge n$.
- Press [ENTER] to see the result, 16.
- Record the result that $a_4 = 16$.

- Next, store the number 10 to n on the Home screen.
- Press [1][0] [STO▸] [X,T,Θ,n].
- Press [ENTER] to see the number10 on the next line.
- Record the result that $a_{10} = 1024$.

Find the 6th and 17th term of the sequence, $a_n = \frac{(-1)^n}{2n}$.

- First store the number 6 to n on the Home screen.
- Press [6] [STO▸] [X,T,Θ,n].
- Press [ENTER] to see the number 6 on the next line.
- Enter the general term of the sequence, $\frac{(-1)^n}{2n}$ as $(-1)^n/(2n)$. Note that the parentheses are essential.
- Press [ENTER] to see the result, .0833333333.
- Press [MATH] and select 1:FRAC→ to convert the result to a fraction. Note that this can all be done on one line.
- Press [ENTER] to see the result, $\frac{1}{12}$.
- Record the result that $a_6 = \frac{1}{12}$.

- Next, store the number 17 to n on the Home screen.
- Press [1] [7] [STO▸] [X,T,Θ,n].
- Press [ENTER] to see the number 17 on the next line.
- Use [2nd] [ENTER] to find the general form of the sequence, or re-enter $\frac{(-1)^n}{2n}$ as $(-1)^n/(2n)$.
- Press [ENTER] to see the result, −.0294117647.
- Press [MATH] and select 1:FRAC→ to convert the result to a fraction.
- Press [ENTER] to see the result, $\frac{-1}{34}$.
- Record the result that $a_{17} = \frac{-1}{34}$.

Calculate n Terms of the Sequence using seq(from the Catalog

Find the first five terms the sequence $a_n = 10 - n^2$.

- From the Home screen, press [2nd] [0] to access the CATALOG.
- The calculator should be in ALPHA mode. There should be an A in the upper right corner of the screen.
- Press the [LN] key to jump to the S's.
- If you do not see an A in the upper right corner of the screen, press the Alpha key first, then the [LN] key to jump to the S's.
- Use the down arrow to find seq(.
- Press [ENTER]. This puts seq(on the Home screen.

- The function ***seq****(expression, variable, begin, end, increment)* returns a list created by evaluating the *expression* with regard to the *variable*, from *begin* to *end* by *increment*.
- Increment may be omitted when the increment is 1.
- Enter the expression, variable, begin, end, increment as $seq(10 - n^2, n, 1, 5)$.
- Press [ENTER] to view the sequence $\{9 \ \ 6 \ \ 1 \ -6 \ -15\}$.

Calculate the Terms of the Sequence from the LIST Menu

Find the first five terms the sequence $a_n = 10 - n^2$.

- To access the seq(function from the LIST menu, first press [2nd] [STAT].
- Move the cursor to the right to go to OPS.
- Move the cursor down to 5:seq and press [ENTER] or press [5].
- This puts seq(on the Home screen.
- Enter the expression, variable, begin, end, increment as $seq(10 - n^2, n, 1, 5)$.
- Press [ENTER] to view the sequence $\{9 \ \ 6 \ \ 1 \ -6 \ -15\}$.

Graph a Sequence

Enter the sequence to be graphed, $a_n = 10 - n^2$

- You must use the DOT mode to display only the points of the sequence. Otherwise, the graph of the sequence looks more like a curve.
- Press [MODE].
- Move the cursor to the word DOT .
- Press [ENTER].
- Note that if you try to use Plot 1, Plot 2 or Plot 3, you get a dimension error screen when you press the [GRAPH] key.

- The basics of graphing a sequence are the same as graphing a function, namely, you must enter the expression that defines the sequence, set the window, etc.
- The settings, however, look very different.
- To graph $a_n = 10 - n^2$, first press the [Y=] key.
- Note that the sequence functions are denoted as *u, v,* and *w*. You may enter three different sequences in a similar way to entering three different functions Y1, Y2, and Y3.

- You must enter a value for the lowest value of *n* you want to use as *n*Min.
- We use *n*Min = 1.
- Next, enter the expression for the sequence next to u_n.
- Use the [X,T,Θ,*n*] key to enter *n*.
- Finally, you must enter the first term of the sequence $u(nMin) = \{9\}$.

Set the Window for the Graph of the Sequence

- Some of the window settings for a sequence are different from the window settings of functions.
- Set *n*Min = 1 and *n*Max = 5, since we want the first five terms of the sequence.
- Set PlotStart = 1 because we want to start at the first term.
- Set PlotStep = 1 because we want to see terms 1, 2, 3, 4, and 5, an increment or step of 1 from term to term.
- Use Xmin = 0 and Xmax = 7 since we want 5 terms.
- The lowest value of the sequence is –15, and the highest is 9, so set Ymin = -20 and Ymax = 15.
- The window settings are shown over two screens.

Graph the sequence

- Press [GRAPH] to see the graph.
- Use [TRACE] to move from point to point.

Create a Table of Values for the sequence $a_n = 10 - n^2$

- To create a table of values of a sequence, enter the general form of the nth term in the Y= menu as described above.
- Press the [Y=] key.
- Note that the sequence functions are denoted as *u, v,* and *w*. You may enter three different sequences in a similar way to entering three different functions Y1, Y2, and Y3.

- You must enter a value for the lowest value of *n* you want to use as *n*Min.
- We use *n*Min = 1.
- Next, enter the expression for the sequence next to u_n.
- Use the [X,T,Θ,*n*] key to enter *n*.
- Finally, you must enter the first term of the sequence $u(nMin) = \{9\}$.

- Press [2nd] [WINDOW] to set up the table.
- Start the table at 1.
- Use increments of 1.
- Set Indpnt: to AUTO.
- Press [2nd] [GRAPH] to see the table.
- Record the first five values of the sequence.

- You may also use the setting Indpnt: ASK and enter only the values of n you want.